Neural Networks and Fuzzy Logic

Neural Networks and Fuzzy Logic

Dr. C. Naga Bhaskar
Principal,
NRI Institute of Technology,
Vijayawada.

Prof. G Vijay Kumar
Department of CSE,
VKR, VNB and AGK College of Engineering,
Gudivada.

BSP **BS Publications**
An Imprint of **BSP Books Pvt. Ltd.**
4-4-309/316, Giriraj Lane,
Sultan Bazar, Hyderabad - 500 095

Neural Networks and Fuzzy Logic
***by* C. Naga Bhaskar & G. Vijay Kumar**

Second Reprint 2025

Published by:

BSP BS Publications
An Imprint of **BSP Books Pvt. Ltd.**
4-4-309/316, Giriraj Lane, Sultan Bazar,
Hyderabad - 500 095
Phone: 040 - 23445688
e-mail: info@bspbooks.net
www.bspbooks.net

Printed at
Repro India Limited.

Price: Rs. 495.00
ISBN: 978-93-81075-40-1 (Paperback)

Preface

Our purpose in writing this book has been to give a systematic account of major concepts and methodologies of artificial neural networks and to present a unified framework that makes the subject more accessible to students and practitioners. This book emphasizes fundamental theoretical aspects of the computational capabilities and learning abilities of artificial neural networks. It integrates important theoretical results on artificial neural networks and uses them to explain a wide range of existing empirical observations and commonly used heuristics.

The main audience are undergraduate students in electrical engineering, computer engineering, and computer science. Alternatively, it may also be used as a valuable resource for practicing engineers, computer scientists, and others involved in research in artificial neural networks.

The background material needed to understand this book is general knowledge of some basic topics in mathematics, such as probability and statistics, differential equations and linear algebra, and something about multivariate calculus. The reader is also assumed to have enough familiarity with the concept of a system and the notion of "state," as well as with the basic elements of Boolean Algebra and Switching Theory. The required technical maturity is that of a senior undergraduate in electrical engineering, computer engineering, or computer science.

Artificial neural networks are viewed here as parallel computational models, with varying degrees of complexity, comprised of densely interconnected adaptive processing units. These networks are fine-grained parallel implementations of nonlinear static or dynamic systems. A very important feature of these networks is their adaptive nature, where "learning by example" replaces traditional "programming" in solving problems. This feature makes such computational models very appealing in application domains where one has little or incomplete understanding of the problem to be solved but where training data is readily available. Another key feature is the intrinsic parallelism that allows for fast computations of solutions when these networks are implemented on parallel digital computers or, ultimately, when implemented in customized hardware.

Artificial neural networks are viable computational models for a wide variey of problems, including pattern classification, speech synthesis and recognition, adaptive interfaces between humans and complex physical systems, function approximation, image data compression, associative memory, clustering, forecasting and prediction, combinatorial optimization, nonlinear system modelling, and control. These networks are "neural" in the sense that they may have been inspired by neuroscience, but not because they are faithful models of biologic neural or cognitive phenomena. In fact, the majority of the network models covered in this book are more closely related to traditional mathematical and/or statistical models such as optimization algorithms, non-parametric pattern classifiers, clustering algorithms, linear and nonlinear filters, and statistical regression models than they are to neurobiological models.

The selection and treatment of material reflect my background as an electrical and computer engineer. The operation of artificial neural networks is viewed as that of nonlinear systems: Static networks are viewed as mapping or static input/output systems, and recurrent networks are viewed as dynamical systems with evolving "state." The systems approach is also evident when it comes to discussing the stability of learning algorithms and recurrent network retrieval dynamics, as well as in the adopted classifications of neural networks as discrete-state or continuous-state and discrete-time or continuous-time. The neural network paradigm and Fuzzy Logic Systems (architectures and their associated learning rules) treated here were selected because of their relevance, mathematical tractability, and/or practicality. Omissions have been made for a number of reasons, including complexity, obscurity, and space.

Every expert system consists of two principal parts: the knowledge base; and the reasoning, or inference, engine.

The *knowledge base* of expert systems contains both factual and heuristic knowledge. *Factual knowledge* is that knowledge of the task domain that is widely shared, typically found in textbooks or journals, and commonly agreed upon by those knowledgeable in the particular field.

Another widely used representation, called the *unit* (also known as *frame*, *schema*, or *list structure*) is based upon a more passive view of knowledge. The unit is an assemblage of associated symbolic knowledge about an entity to be represented. Typically, a unit consists of a list of properties of the entity and associated values for those properties.

Since every task domain consists of many entities that stand in various relations, the properties can also be used to specify relations, and the values of these properties are the names of other units that are linked according to the relations. One unit can also represent knowledge that is a "special case" of another unit, or some units can be "parts of" another unit.

The *problem-solving model,* or *paradigm,* organizes and controls the steps taken to solve the problem. One common but powerful paradigm involves chaining of IF-THEN rules to form a line of reasoning. If the chaining starts from a set of conditions and moves toward some conclusion, the method is called *forward chaining.* If the conclusion is known (for example, a goal to be achieved) but the path to that conclusion is not known, then reasoning backwards is called for, and the method is *backward chaining.* These problem-solving methods are built into program modules called *inference engines* or *inference procedures* that manipulate and use knowledge in the knowledge base to form a line of reasoning.

The *knowledge base* an expert uses, is what he learned at school, from colleagues, and from years of experience. Presumably the more experience he has, the larger his store of knowledge. Knowledge allows him to interpret the information in databases advantage with of diagnosis, design, and analysis.

Though an expert system consists primarily of a knowledge base and an inference engine, a couple of other features are worth mentioning: reasoning with uncertainty, and explanation of the line of reasoning.

Knowledge is almost always incomplete and uncertain. To deal with uncertain knowledge, a rule may have associated with it a *confidence factor* or a weight. The set of methods for using uncertain knowledge in combination with uncertain data in the reasoning process is called *reasoning with uncertainty*. An important sub-class of methods for reasoning with uncertainty is called "fuzzy logic," and the systems that use them are known as "fuzzy systems."

-Authors

Acknowledgements

First and foremost, we acknowledge the contributions of the many researchers in the area of artificial neural networks on which most of the material in this text is based. It would have been extremely difficult (if not impossible) to write this book without the support and assistance of a number of organizations and individuals

We thank our students, who have made classroom use of preliminary versions of this book and whose questions and comments have definitely enhanced it. In particular, the faculty of CSE, IT and EEE, who reviewed the book. We also would like to thank many colleagues in the artificial neural networks for many enjoyable and productive conversations and collaborations.

We are indebted to Vasudeva Rao, BS Publications, who capably and enthusiastically supported in completing the manuscript.

Our deep gratitude goes to the reviewers for their critical and constructive suggestions. They are Professors Dr G N Swamy, Electronics Department, Gudlavalleru Engineering College, Dr G V Raju, Sri Sunflower College of Engineering and Technology, and other anonymous reviewers.

Finally, let me thank the Management of VKR, VNB and AGK College of Engineering, who extended the morale and technical support and for their quiet patience through the many hours during the preparation of the manuscript.

-Authors

Contents

Chapter 8B

Fuzzy Logic Application

CHAPTER 1

OVERVIEW OF NEURAL NETWORKS

Introduction

Throughout the years, the computational changes have brought growth to new technologies. Such is the case of artificial neural networks, that over the years, they have given various solutions to the industry.

Designing and implementing intelligent systems **has become a crucial factor for the innovation and development of better products for society**. Such is the case of the implementation of **artificial life** as well as giving solution to interrogatives that linear systems are not able resolve.

A neural network is a parallel system, capable of resolving paradigms that linear computing cannot. A particular case is for considering which I will cite. During summer of 2006, an intelligent crop protection system was required by the government. This system would protect a crop field from season plagues. The system consisted on a flying vehicle that would inspect crop fields by flying over them.

Now, imagine how difficult this was. Anyone that could understand such a task would say that this project was designated to a multimillionaire enterprise capable of develop such technology. Nevertheless, it wasn't like that. The selected company was a small group of graduated engineers. Regardless the lack of experience, the team was qualified. The team was divided into 4 sections in which each section was designed to develop specific sub-systems. The leader was an electronics specialist. She developed the electronic system. Another member was a mechanics and hydraulics specialist. He developed the drive system. The third member was a system engineer who developed all software, and the communication system. The last member was designed to develop all related to avionics and artificial intelligence.

Everything was going fine. When time came to put the pieces together, all fitted perfectly until they find out the robot had no knowledge about its task. What happened? The one designated to develop all artificial intelligent forgot to "teach the system". The solution would be easy; however, training a neural network required additional tools. The engineer designated to develop the intelligent system passed over this inconvenient.

History of Neural Networks

The study of the human brain dates back thousands of years. But it has only been with the dawn of modern day electronics that man has begun to try and emulate the human brain and its thinking processes. The modern era of neural network research is credited with the work done by neuro-physiologist, Warren McCulloch and young mathematical prodigy Walter Pitts in 1943. McCulloch had spent 20 years of life thinking about the "event" in the nervous system that allowed to us to think, feel, etc. It was only until the two joined forces that they wrote a paper on how neurons might work, and they designed and built a primitive artificial neural network using simple electric circuits. They are credited with the McCulloch-Pitts Theory of Formal Neural Networks. (Haykin, 1994, pg: 36) (http://www.helsinki.fi)

The next major development in neural network technology arrived in 1949 with a book, "The Organization of Behavior" written by Donald Hebb. The book supported and further reinforced McCulloch-Pitts's theory about neurons and how they work. A major point brought forward in the book described how neural pathways are strengthened each time they were used. As we shall see, this is true of neural networks, specifically in training a network. (Haykin, 1994, pg: 37)(http://www.dacs.dtic.mil)

During the 1950's traditional computing began, and as it did, it left research into neural networks in the dark. However certain individuals continued research into neural networks. In 1954 Marvin Minsky wrote a doctorate thesis, "Theory of Neural-Analog Reinforcement Systems and its Application to the Brain-Model Problem", which was concerned with the research into neural networks. He also published a scientific paper entitled, "Steps Towards Artificial Intelligence" which was one of the first papers to discuss Artificial Intelligence in detail. The paper also contained a large section on what nowadays is known as neural networks. In 1956 the Dartmouth Summer Research Project on Artificial Intelligence began researching Artificial Intelligence, what was to be the primitive beginnings of neural network research. (http://www.dacs.dtic.mil)

Years later, John von Neumann thought of imitating simplistic neuron functions by using telegraph relays or vacuum tubes. This led to the invention of the von Neumann machine. About 15 years after the publication of McCulloch and Pitt's pioneer paper, a new approach to the area of neural network research was introduced. In 1958 Frank Rosenblatt, a neuro-biologist at Cornell University began working on the Perceptron. The perceptron was the first "practical" artificial neural network. It was built using the somewhat primitive and "ancient" hardware of that time. The perceptron is based on research done on a fly's eye. The processing which tells a fly to flee when danger is near is done in the eye. One major downfall of the perceptron was that it had limited capabilities and this was proven by Marvin Minsky and Seymour Papert's book of 1969 entitled, "Perceptrons". (http://www.dacs.dtic.mil) (Masters, 1993, pg: 4-6)

Between 1959 and 1960, Bernard Wildrow and Marcian Hoff of Stanford University, in the USA developed the ADALINE (ADAptive LINear Elements) and MADALINE (Multiple ADAptive LINear Elements) models. These were the first neural networks that could be applied to real problems. The ADALINE model is used as a filter to remove echoes from telephone lines. The capabilities of these models were again proven limited by Minsky and Papert (1969). (http://www.dacs.dtic.mil).

The period between 1969 and 1981 resulted in much attention towards neural networks. The capabilities of artificial neural networks were completely blown out of proportion by writers and producers of books and movies. People believed that such neural networks could do anything, resulting in disappointment when people realized that this was not so. Asimov's television series on robots highlighted humanity's fears of robot domination as well as the moral and social implications if machines could do mankind's work. Writers of best-selling novels like "Space Oddesy 2001" created fictional sinister computers. These factors contributed to large-scale critique of Artificial Intelligence and neural networks, and thus funding for research projects came to a near halt. (Haykin, 1994, pg: 38) (http://www.dacs.dtic.mil)

An important aspect that did come forward in the 1970's was that of self-organizing maps (SOM's). Self-organizing maps will be discussed later in this project (Haykin, 1994, pg: 39). In 1982 John Hopfield of Caltech presented a paper to the scientific community in which he stated that the approach to Artificial Intelligence should not be to purely imitate the human brain but instead to use its concepts to build machines that could solve dynamic problems. He showed what such networks were capable of and how they would work. It was his articulate, likeable character and his vast knowledge of mathematical analysis that convinced scientists and researchers at the National Academy of Sciences to renew interest into the research of Artificial Intelligence and neural networks. His ideas gave birth to a new class of neural networks that over time became known as the Hopfield Model (http://www.dacs.dtic.mil) (Haykin, 1994, pg: 39).

At about the same time at a conference in Japan about neural networks, Japan announced that they had again begun exploring the possibilities of neural networks. The United States feared that they would be left behind in terms of research and technology and almost immediately began funding for AI and neural network projects (http://www.dacs.dtic.mil).

1986 saw the first annual Neural Networks for Computing conference that drew more than 1800 delegates. In 1986 Rumelhart, Hinton and Williams reported back on the developments of the back-propagation algorithm. The paper discussed how back-propagation learning had emerged as the most popular learning set for the training of multi-layer perceptrons. With the dawn of the 1990's and the technological era, many advances into the research and development of artificial neural networks are occurring all over the world. Nature itself is living proof that neural networks do in actual fact work. The challenge today lies in finding ways to electronically implement the principles of neural network technology. Electronics companies are working on three types of neuro-chips namely, digital, analog, and optical. With the prospect that these chips may be implemented in neural network design, the future of neural network technology looks very promising.

1.1 BIOLOGICAL VS. ELECTRICAL BRAINS

Biological and electrical brains are very similar in some aspects and very different in others. One of the main similarities is the modeling of neurons. A biological brain is a collection of individual neurons that send electric pulses to one another based on reactions and pulses perceived. An electrical brain is very similar in which "nodes" send electrical signals to one another through electric wires. These pulses then enact different responses in the various neurons influenced by them. A minor difference does lie in the pulses though, while a biological brain can vary the electric pulse in amplitude, most electric brains are stuck inside a specific voltage range.

The big difference between a biological brain and an electric brain is the ability of the biological brain to radically alter its structure while learning. A human brain on the other hand, learns by re-arranging the structure of the brain. A neuron in a human brain can alter its paths and electric charge to affect those around it. On the other hand, an electric brain must store information and give certain paths different weights. Paths never change or disappear, inside they grow stronger or weaker. Simply put, a biological brain's hardware can change, while neural networks hardware cannot. Learning in general is a very hard concept for the electrically brain to grasp. This occurs because a set of guidelines and rules must be laid out for the electronic brains so it understands what must be remember and what must be committed to memory.

1.1.1 LAYERS

There are 3 main layers in a neural network. The first of these layers is the input layer. In the input layer, data is gathered from external sources. These external sources can be sensors, manually inputted data, or data generated by other neural networks or the same network. The input layer then passes the data to a hidden layer. Because the input layer accepts data, it often acts as a buffer for the hidden layer. The hidden layer serves two functions. The first function is processing the data. Here equations are solved, or answers are formulated. The second function of the hidden layer is to determine what is learned and what is forgotten. Here the learning rules and laws are applied, and the "structure" of the neural network is updated. Lastly, there is the output layer. The output layer is where the processing and data meet the external world. The output layer could be a set of lights, or a computer screen, even a voice synthesizer. This layer just like the input layer is one way. This one way creates a buffer that can protect the more sensitive hidden layers.

1.1.2 COMMUNICATIONS

An important aspect in a neural network is how the neurons communicate with one another. There are three different types of communication. The first is inter-layer connections. These are the communication lines used to communicate this from one layer to the next. When using this type of communication, the sending layer cannot vary by from the receiving layer by more than a difference of one. In other words, this means layer 1 can talk to only layer 0 or 2, using inter-layer communications.

The next type of communications is the intra-layer communications. This is where neurons within the same layer communicate with one another. Besides, neurons talking to other neurons, there are self-connections, which are when a neuron talks to itself. These connections are considered a special type of intra-layer communications. Lastly, there is the supra layer communications. These are the communications that occur when a neuron needs to talk to another neuron or layer that is more than a difference of 1 away. For example, layer 1 sending a message to layer 5, would be considered a supra layer communication. All three of these communications are required for a successful neural network.

Communications between neurons and layers are often weighted.

1.1.3 INTER-LAYER

Taking a more in-depth look at inter-layer communication, it can be seen that there are two different types of connections. The first type of connection is the full connection. A full connection is one that tries to maximize the number of connections between neurons. A specific definition cannot be developed, due to the fact that learning methods influence, which connections are valid and which ones are not. Three variations are possible with full connections. The most common is the fully interlayer-connected network. This communication method is where all possible connections are present between layers, but no intra or supra layer connections exist (Figure 1.2). Another method, which many people think of when fully connected is mentioned is the plenary neural network. This network communication method has all possible connections, including those found in the intra and supra layers (Figure 1.2). Lastly, the third method is a plenary network that doesn't employ self-connections. This reduction in self-connections, increases the speed of the network, but does have an impact on the ability of neurons to learn.

1.2 INTRODUCTION TO BIOLOGICAL NEURON

Artificial Neural Networks (ANNs) are computer systems (software[1] or hardware) that are biologically inspired in that they attempt to simulate the processing capabilities of the networks of **neurons** in the human brain.

The number of applications of ANNs to real world problems is immense and they permeate industry and commerce. ANNs perform particularly well where there is a large amount of historical data, where the application involves recognizing patterns in the data or where the problem is one of classification. ANNs fit into the general area of **computational intelligence**[2] and rank alongside **fuzzy logic** as the most successful.

The biological inspiration for ANNs is motivated by the fact that the human brain is capable of so much when compared with a computer. Although computers can very quickly process numbers and carry out complex mathematical calculations they are unable to reason in a similar manner to human beings. So, the suggestion is that we should simulate some of the physical processes that provide the ability to reason and tackle difficult problems that computers allied to mathematical techniques are unable to solve well. The basic premise is that we should borrow from nature.

The eventual aim is to emulate the way human beings think. This is still only a pipe dream but currently ANNs are able to tackle complex real world problems that mathematical and statistical techniques are often unable to handle. In particular ANNs are able to deal with 'noisy' data and have strong generalisation capabilities. Noisy data is real world data that does not, for

---

[1] The vast majority of applications are software implementations of ANNs.

[2] Computational Intelligence is a term used to cover three techniques – artificial neural networks, fuzzy logic and genetic algorithms – that attempt to imbue computers with some 'intelligence'.

example, fit exactly to a mathematical function but contains random variations. Generalisation is the ability to handle examples that the network hasn't seen before. We are still unsure exactly how the brain operates but we understand enough to enable us to mimic the basic operations and interactions between neurons. The human brain is made up of approximately 1×10^{11} neurons with of the order of 1×10^{15} connections between them. Figure 1.1 provides a schematic diagram for a biological neuron and the connections between neurons.

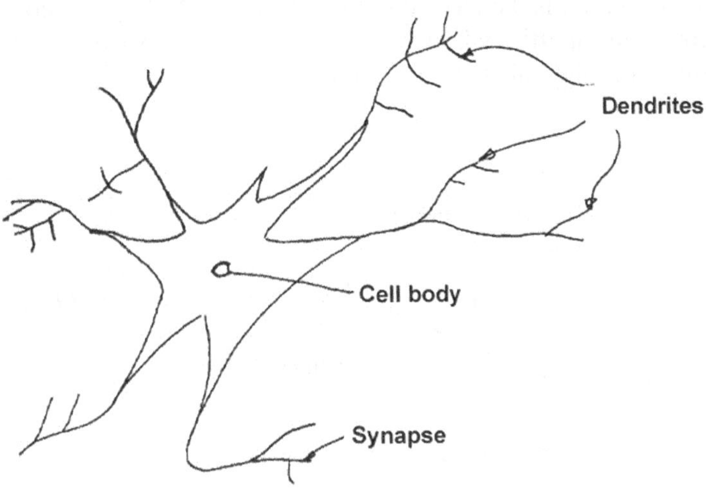

Figure 1.1 The Biological Neuron (Wasserman, 1989).

Each neuron in the brain possesses the capability to take electrochemical signals, as input, and process them before sending new signals via the connections between neurons, known as the dendrites. The cell body receives these signals at the synapse. When the cell body receives the signals they are summed – some signals excite the cells whilst others will inhibit the cell. On exceeding a threshold, a signal is sent via the dendrites to other cells. It is this receiving of signals and summation procedure that is emulated by the artificial neuron.

The biological neuron is only an inspiration for the artificial neuron. The artificial neuron is clearly a simplification of the way the biological neuron operates and indeed much is still not known about the way the brain operates for us to carry the analogy too far. ANNs, then, are an approach for modelling a simplification of the biological neuron – we are not modelling the brain but merely using it as an inspiration.

We can summarise this section in the following way.

- ANNs learn from data, are biologically inspired and are based on a network of artificial neurons;
- ANNs handle noisy data and are able to generalise in that they can cope with examples of problems that they have not experienced before.

1.3 ARTIFICIAL NEURON

The artificial neuron is the basic building block of ANNs (from now on we will use neuron to mean artificial neuron). There are a number of variations on this basic neuron but they all have the same simple design. Figure 2.2 shows the basic structure.

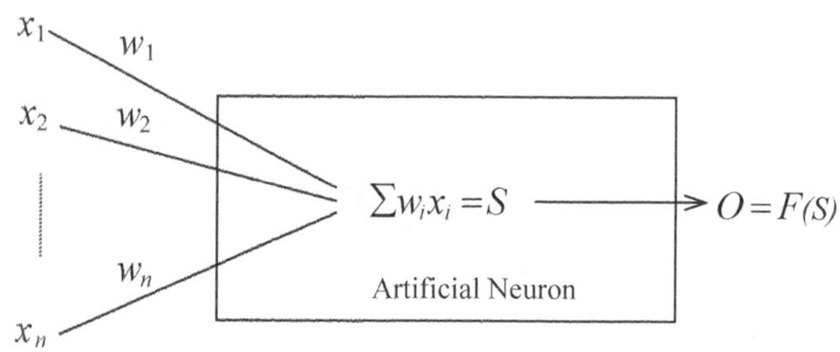

Figure 1.2 The Basic Artificial Neuron.

Each neuron receives inputs, $x_1, x_2, \ldots, x_n$, which are connected to the input side of the neuron. Attached to every connection is a weight w_i which represents the connection strength for that input.

The cell node then calculates the weighted sum of the inputs given by

$$S = \sum_{i=1}^{n} w_i x_i$$

An **activation function,** F, takes the signal, S, as input to produce the output, O, of the neuron. In other words

$$O = F(S)$$

There are a number of functions that could be employed. For example it may simply be a threshold

$$O = \begin{cases} 1 & \text{where } S > T \\ 0 & \text{otherwise} \end{cases} \qquad \text{.....(1.1)}$$

where T is a constant threshold. This can be interpreted to mean that if the weighted sum is above T then the node 'fires'.

Another commonly adopted activation function is the logistic or sigmoidal function given by

$$O = \frac{1}{1+e^{-S}} \quad (-\infty < S < \infty) \qquad .....(1.2)$$

This has the effect of limiting the output of the neuron to a minimum of zero and a maximum of S. Figure 2.3 shows the effect of applying the sigmoidal function.

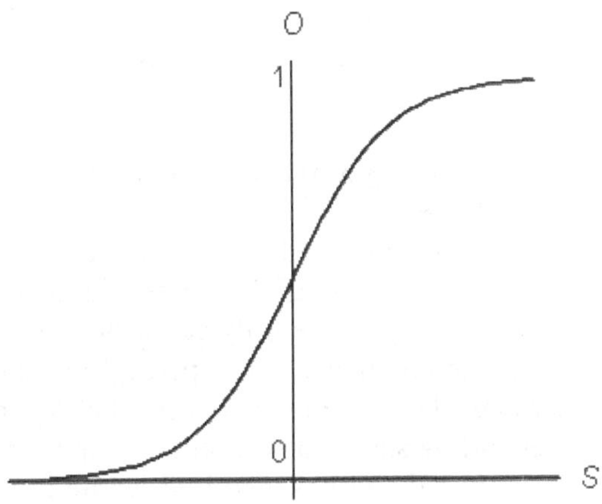

Figure 1.3 The Sigmoidal Activation Function.

As you can see, this has the effect of compressing the value of O to between zero and one. This function also has the effect of introducing nonlinearity into the network.

There are a number of other functions that are employed in real applications. Choosing an activation function is one of the many decisions faced by a neural network developer.

In this basic description of a neuron, we have already come across the notion of a weight. It's not too simplistic to say that the problem of developing an ANN is primarily to find a method for learning or estimating these weights. You will see this more clearly later.

1.3.1 THE PERCEPTRON

Neural networks research started in the 1940s when McCulloch and Pitts (1943) introduced the idea of the perceptron. The original perceptron consisted of a single **layer** of neurons that had a threshold activation function as described above. An example is shown in Figure 1.4.

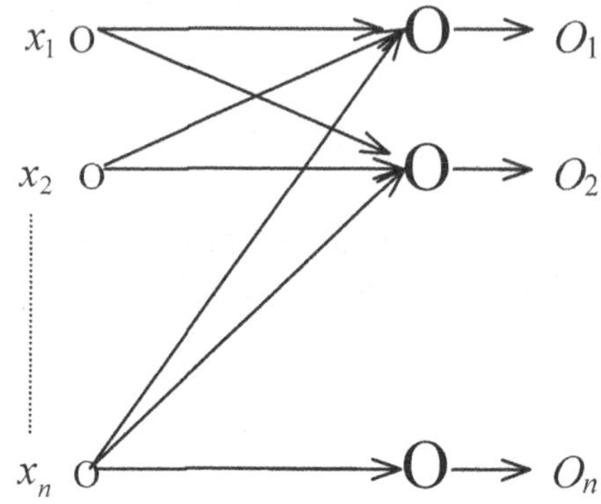

Figure 1.4 The Perceptron.

The input layer feeds the data through to the output layer where the neurons process the inputs and weights using the threshold activation function as described by eq. 1.1. The two layers are fully connected in that each input node is connected to each output node. The perceptron carries out pattern classification by learning the weights between the layers by **supervised learning** where the network is supplied with known input and output data. The differences between supervised and unsupervised learning will be discussed in Section 4.

A typical problem that was tackled in the early days of the perceptron is where the input and output data is binary (i.e. 0 or 1) and we have a set of inputs and outputs. This is an example of a training data set and each pair of inputs and outputs is known as a pair of training patterns. The perceptron learns the weights in the following way.

1. The weights are randomly initialised to small values.
2. For the first input pattern the network output is calculated using eq. 1.1. Denote this output by O_{jp} for input pattern p and output unit j.
3. The error is calculated as $T_{jp} - O_{jp}$ where T_{jp} represents the known output.
4. A simple learning rule for the weights might be

$$w_{ji}^{new} = w_{ji} + c(T_{jp} - O_{jp})a_i$$

where

c is a constant ($0 \leq c \leq 1$) representing the learning rate

a_i is the value of input unit i

w_{ji} is the weight from node j to node i

w_{ji}^{new} is the new weight from node j to node i.

5. The process is repeated for the next input pattern.

This process is repeated until there is an acceptable error in the classification, where the error is usually a root-mean-square measure such as

$$\frac{\sqrt{\sum_{p=1}^{n_p}\sum_{j=1}^{n_o}(T_{jp}-O_{jp})^2}}{n_p n_o}$$

where n_p is the number of patterns in the training set and n_o is the number of output units.

The exciting fact about this approach is that it was proved that this algorithm could learn anything it could represent. In other words the weights could be adjusted to simulate any function that could be represented by inputs and outputs. However there was a problem. In 1969 Professor Minsky found a whole class of problems that a perceptron could not simulate – problems that are not linearly separable. The best example of this is the exclusive-OR function. This is where we have two inputs, x and y, and an output, out; the relationships are given in Table 1.1 and the graphical representation is in Figure 2.5.

Table 1.1 The Exclusive-OR Problem

x	y	Out
0	0	0
1	0	1
0	1	1
1	1	0

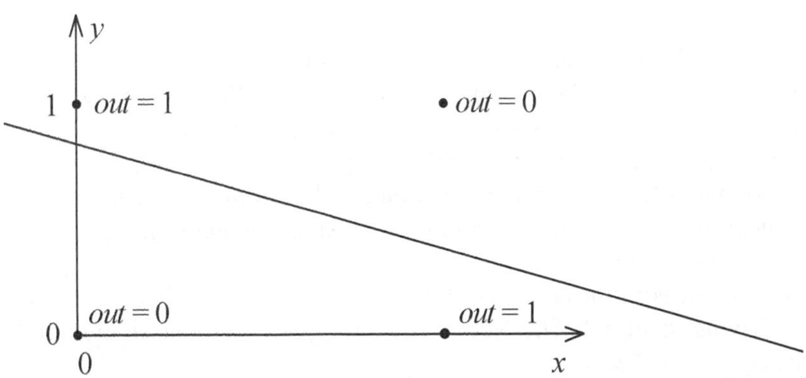

Figure 1.5 Graphical Representation of the Exclusive-OR Problem.

For a single neuron with threshold activation function with two inputs and one output it can be shown that there are no two weights that will produce a solution. It is not linearly separable in that there is no straight line which will 'split' the points with output values of 1 from those with output values of 0. Unfortunately, it seems that most real world problems are not linearly separable. This caused the development of neural networks to be put on hold until Rumelhart and McClelland (1986) put forward the notion of the **multilayered perceptron** with three layers of neurons. These networks form the basis for much of the success of neural networks today. In particular, multilayered neural networks that use the **backpropagation** algorithm (see Section 2.3) are employed in most successful ANN applications.

1.4 THE HODGKIN-HUXLEY MODEL

It is evident that Hodgkin and Huxley's theory and mathematical model for the generation of the action potential, published in complete form in 1952, was widely admired and received careful consideration, as shown in several references. Some of the historical and elementary theoretical background will be reviewed here. The intent is to present a simple, clear picture of the action potential and its associated refractory periods, for use later in studying the dynamics of small reverberatory networks. Perhaps this point of view will also be helpful to others, such as retired engineers, who would like to contribute to neuroscience as amateurs.

Hodgkin and Huxley's work was an experimental and theoretical triumph, but it developed out of a gradually clarifying view of neural processes, to which many people contributed. Some of these are mentioned by Hille (1984, pp. 24-28).

1.4.1 RESTING POTENTIAL

Bernstein used Walter Nernst's equation, which is based on thermodynamic principles, to calculate the resting potential inside the membrane from the concentration gradient for potassium. The qualitative explanation of this effect is that the membrane is semipermeable to potassium (but supposedly not to sodium) so that potassium tends to leak out, and in doing so the potassium ions leave a negative charge in the cytoplasm inside the neuron. This sets up a potential difference between inside and outside, which tends to drive potassium ions back in. An equilibrium is established between this and the concentration gradient driving them out. The Nernst equation gives the potassium equilibrium potential as

$$E_K = (RT/zF) \ln ([K^+]_o/[K^+]_i)$$

where R and F are physical constants, z is valence, and T is the absolute temperature (degrees K). $[K^+]_o$ is the outer concentration of K^+, and $[K^+]_i$ is the inner concentration.

1.4.2 The Equivalent Circuit

To facilitate complete description of the factors determining resting potential in neurons, an electrical equivalent circuit can be used. This circuit represents a short cylindrical section of an axon, over which the membrane potential is approximately uniform. The equivalent circuit is shown in Fig. 1.6. The potential at one node, labeled V_m, corresponds to the uniform potential. The membrane has the distributed properties of conductance (the inverse of resistance) and capacitance, both of which are represented by discrete elements in the circuit. Biologically, there is a mechanism called the Na^+-K^+ pump, which metabolically establishes concentration gradients for sodium and potassium ions across the membrane. This is not represented by a circuit element; it simply maintains approximately fixed concentrations for sodium and potassium ions.

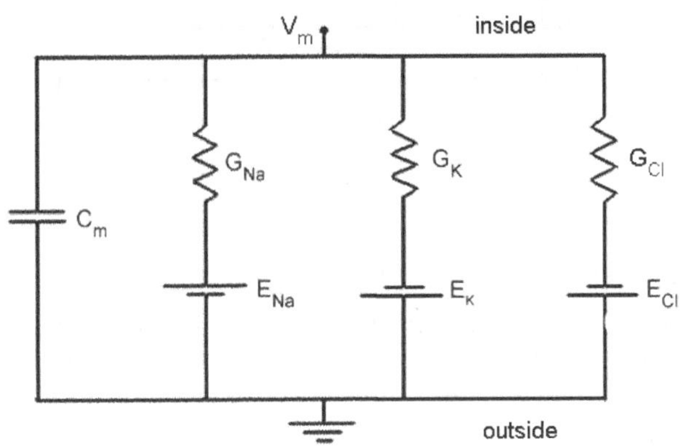

Figure 1.6 The conductances G_{Na}, G_K, and G_{Cl} (Cl for chloride) are small, so that the axon membrane is approximately an insulator separating two conductors: the cytoplasm and the external fluid. The cytoplasm is represented by the node with the membrane potential V_m. The external fluid corresponds to the lower node connected to the ground symbol. The ground indicates that the potential of the external fluid is being taken as the reference potential, and is therefore zero.

1.4.3 The Resting Conductance

The conductances G_{Na}, G_K, and G_{Cl} are total values representing a great many individual ion channels. They are constants during the resting potential. Later, in modeling the action potential, G_{Na} and G_K will be taken as variables dependent on V_m and time. The individual ion channels are diverse as well as numerous. Studies of the genes that encode ion channels indicate that a great many combinations of properties are possible, so that channels can serve a variety of special purposes. For the study of membrane potential in axons, ion channels can be divided into resting channels, which are always open, and voltage-gated channels which only open during the action potential. Thus, in the discussion of the resting potential, the symbols G_{Na}, G_K, and G_{Cl} only represent the totals of the resting channels of the three ionic types in Fig. 1.7. The three can be added together to give the total resting membrane conductance G_M.

$$G_M = 1/3333.3 = 3.0000 \times 10^{-4} \text{ mho/cm}^2 = 0.30000 \text{ mmho/cm}^2$$

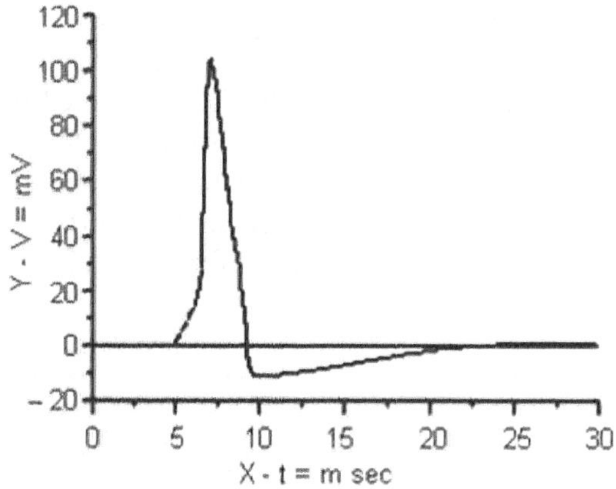

Figure 1.7 Resting Potential.

1.4.4 THE ACTION POTENTIAL

An action potential is shown in Fig. 1.8. This is taken from the Squid tutorial in Bower and Beeman (1998), which represents a section of axon with membrane properties similar to those of the squid giant axon. The figure is only part of a window in the Squid display. It can be seen that after the spike there is an after-potential which drops below 0 mV (This is the resting potential, taken as zero to conform with Hodgkin and Huxley; considered – 70 mV elsewhere) before settling down. This is the time interval in which there is a refractory period.

Hodgkin and Huxley's outstanding achievement was the quantitative explanation of the action potential. Their work was based on the sodium hypothesis presented by Hodgkin and Katz (1949). This proposed that, during activity after stimulation above the threshold, the membrane becomes selectively highly permeable to sodium ions, so that the inward flow of sodium through the membrane briefly eclipses the outward flow of potassium, and the internal membrane potential goes positive past zero (103 – 70 = +33 mV), and rises toward the sodium equilibrium potential. Then the sodium permeability declines and the potassium permeability becomes large, until the spike is completed.

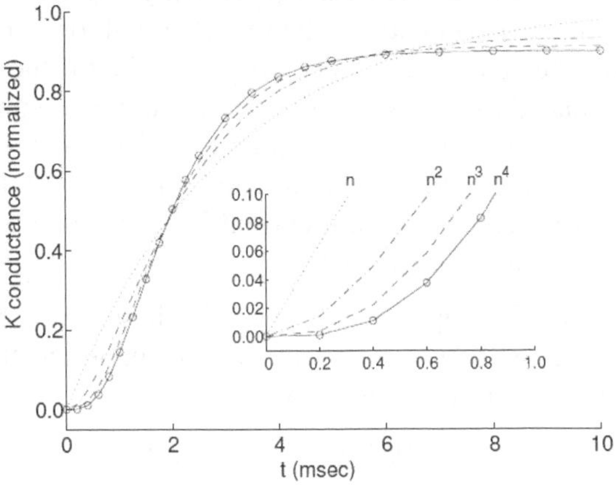

Fig. 1.8 The initial inflection in the curve cannot be well fit by a simple exponential (dotted line) that rises linearly from zero. Successively higher powers of p (p =2: dot-dashed; p = 3: dashed line) result in a better fit to the initial inflection. In this case, p = 4 (solid line) gives the best fit

For quantitative application of the sodium hypothesis to calculate the action potential, values for the sodium and potassium conductances as functions of voltage and time were needed. Hodgkin and Huxley obtained these by means of a voltage-clamp technique, which is described by Johnston and Wu (1995) on pages 143-145. Two silver-wire electrodes were inserted axially through one end of a length of axon. One electrode measured the membrane potential V_m, and a circuit compared it with a command voltage which had been set by the experimenter. If V_m deviated from the command voltage, a current was injected at the second electrode to eliminate the deviation. Thus V_m was clamped to the command value.

The experiments started from a holding voltage which was apparently intended to equal the resting potential. Hodgkin and Huxley (1952a), page 455 mention estimates of – 60 to – 65 mV for the actual resting potential, depending on factors such as junction potential, which modify the measured value (Their signs are reversed, being given as + 60 to + 65, conforming to the choice mentioned in Hodgkin, Huxley, and Katz (1952), page 425). A range of values for the command voltage was used. Following the account in Johnston and Wu, the starts were from a holding voltage of – 60 mV relative to the external fluid, followed by a voltage step to a command voltage, which was then maintained for 8 msec.

The injected current required to maintain the command voltage was recorded as a function of time. The injected currents used in the above procedure pass through the membrane to return to the external circuit. Thus they are equal to sums of the sodium and potassium currents flowing through the membrane. To separate the twos, the experiments can be repeated with one or the other type of channels blocked. The sodium channels will be blocked if the extracellular Na^+ concentration is reduced to its inside value, so that the Na^+ equilibrium potential is zero. They can also be blocked with tetrodotoxin. The potassium channels can be blocked by removing intracellular K^+, or with tetraethylammonium. The result is a family of curves for sodium current I_{Na} as a function of time with V_m as parameter, and a similar family for I_K.

Johnston and Wu (1995, pp. 145-147) describe how it was demonstrated that the instantaneous relation between current and voltage in the above experiments is linear; that is, that it follows Ohm's law, so that

$$I_{Na} = G_{Na} (V_m - E_{Na})$$

and

$$I_K = G_K (V_m - E_K).$$

Thus the conductances G_{Na} and G_K can be obtained as functions of V_m and time t by dividing the respective currents by their driving voltages.

The foregoing procedures give G_{Na} and G_K as functions of V_m and t over the range of discrete values of V_m and t which were tested. For use in simulation calculations of the action potential, the discrete data points must be fitted with continuous functions. Hodgkin and Huxley tried to relate the complex curve forms at particular values of V_m to exponentials, and in doing so were led to express them in terms of maximum values of G_{Na} and G_K which are multiplied by coefficients ranging between zero and 1. The coefficients turned out to be small integral powers of exponentials. The theoretical significance of these forms is discussed in Johnston and Wu (1995, pp. 149-154), and in Bower and Beeman (1998, pp. 37-38). The equations for approximating the voltage dependence are described in Bower and Beeman (1998, pp. 44-47).

Having constructed expressions for calculating the sodium and potassium conductances, Hodgkin and Huxley proceeded to use them to calculate the action potential. It was 1951, and the early Cambridge University computer was down for about 6 months for modification. Huxley managed to carry out numerical integrations of the differential equations with a hand-operated mechanical calculator. They were excited to see the action potential come out with the right shape. They saw that explanation of the action potential in the squid axon in terms of sodium and potassium conductances is fundamental, and that therefore it might be expected that similar explanations would apply to other excitable tissues, perhaps with different values for parameters.

1.4.5 COMPUTER SIMULATION OF NEURAL BEHAVIOR

The Hodgkin-Huxley theory provides a foundation for modeling neurons. It is especially suited to simulations on digital computers, since they handle the numerical integrations easily. The fundamental character of the theory gives assurance of a close correspondence to biological reality, and this gives the simulations value as natural experiments.

The GENESIS computer program for computational neuroscience (Bower and Beeman, 1998) is a flexible system for simulation, providing building blocks for simulations both below and above the level of the whole neuron. With GENESIS scripting, users can also design their own extended building blocks. The GENESIS tutorial named Squid, which is flexible in itself, embodies the classical Hodgkin-Huxley model of the giant axon. The Squid tutorial is of special interest here, because the chapter in Bower and Beeman (1998) which covers it includes Exercise 7 (p. 49) for investigating the refractory periods.

1.5 WHY SPIKING NEURONS?

Models of spiking neurons have been called the 3rd generation of artificial neural network (Maass 1997), as in

- Generation 1: Binary networks (activation of 0 or 1) such as implemented by McCulloch and Pitts' neurons and the Hopfield network.
- Generation 2: Real-valued networks, where activation is representative of the 'mean firing rate' of a neuron, such as Backpropagation networks and Kohonen self-organising maps.
- Generation 3: Spiking neural networks (SNNs).

Networks of the earlier generations have proven effective at modelling some cognitive processes and have been successful in many engineering applications. However the fidelity of these models with regards to neurophysiological data is minimal and this has several drawbacks.

- Neurophysiological knowledge cannot be integrated easily into the models and as such cannot be tested for applicability to or effect upon neural computation.
- Real neurons exhibit a very broad range of behaviours (tonic (continuous) and phasic (once-off) spiking, bursting, spike latency, spike frequency adaptation, resonance, threshold variability, input accommodation and bistability (Izhikevich 2004)). It's unlikely that these behaviours have no computational significance.
- There are specific interesting processes occurring at the spike level (such as Spike Timing Dependent Plasticity (Bi and Poo 1998)) that cannot be modelled without spikes.
- The dynamics of spiking networks are much richer, allowing for example

- oscillations in network activity which could implement multiple concurrent processing streams (Izhikevich 1999), figure/ground segmentation and binding (Csibra, Davis et al. 2000; Engel, Fries et al. 2001), short term memory (Jensen, Idiart et al. 1996; Jensen, Gelfand et al. 2002) etc
- much increased (perhaps by orders of magnitude) memory capacity (Izhikevich 2005). Transmission delays are very significant for computation particularly because they are random or Gaussian for real neurons - this causes the formation of polychronous (as against synchronous) spiking neuron groups which could possibly store many more population-encoded memories than there are synaptic weights. This idea is still to be fully researched and analysed.

One of the most exciting characteristics of spiking neural networks, with the potential to create a step-change in our knowledge of neural computation, is that they are innately embedded in time (Maass 2001). Spike latencies, axonal conduction delays, refractory periods, neuron resonance and network oscillations all give rise to an intrinsic ability to process time-varying data in a more natural and computationally powerful way than is available to 2nd generation models. Real brains are embedded in a time-varying environment; almost all real-world data and human or animal mental processing has a temporal dimension. Evidence is growing that rhythmic brain oscillations are strongly connected to cognitive processing (Klimesch 1999; Basar, Basar-Eroglu et al. 2001; Engel, Fries et al. 2001; Kahana, Seelig et al. 2001; Ward 2003). So utilising Spiking Neural Networks may be one of the first steps needed to bridge the current divide between existing ANN models and more flexible, realistic and, dare I say, intelligent, behaviour from artificial systems.

1.5.1 POTENTIAL DISADVANTAGES

- **Processing time**: Until recently, biologically realistic spiking models have required intensive computations for even small amounts of simulated time, making simulations of large networks or long time periods impractical in most situations. However this is no longer of such great concern with the formulation of a model which is both biologically realistic and computationally efficient (see the next section: What spiking neuron models are available?)
- **Complex dynamics**: One of the advantages may also be a disadvantage in that the complex behaviour needs to be understood and effectively managed. Although there are analogs to basins of attraction for networks of spiking neurons on a global scale as long as noise is present in the system (Gerstner 1995), there is no such thing as a 'stable state' on a local neuron scale unless the entire network is completely inactive, so the network is always dynamically changing at some level.

- **Limited knowledge**: Much less is known about networks of spiking neurons than the more established ANN paradigms, and many well-accepted methodologies need to be adapted or possibly replaced. For instance with the right choice of parameters, STDP appears to be able to intrinsically support Hebbian learning with no need for arbitrary weight bounds or explicit weight or firing rate normalisation (van Rossum, Bi et al. 2000).

1.5.2 WHAT SPIKING NEURON MODELS ARE AVAILABLE?

Models fall into 3 broad categories ranging from complex to simple.

1. Systems of coupled equations of 2 or more variables using parameters with real biophysical correlates.
2. Systems of coupled equations of 2 or more variables using parameters with **no** biophysical correlates.
3. Integrate and Fire models.

These categories are not entirely mutually exclusive; some overlap is evident. Discussion of each follows.

1.5.3 COUPLED EQUATIONS USING PARAMETERS WITH REAL BIOPHYSICAL CORRELATES

Although it was published more than 50 years ago, the definitive model here remains the Hodgkin-Huxley model (Hodgkin and Huxley 1952). Its advantage is that it completely describes neuron behaviour in terms of all known biophysical parameters and as such is theoretically able to model any possible neuron behaviour. In practice, the parameter values can be difficult to determine, and in fact the values required to successfully emulate several known neuron types using this model are yet to be found (Izhikevich 2004). The model also has excessive computational requirements and hence cannot be used to simulate large networks over long time scales.

There are several simplified HH models available (Morris and Lecar 1981; Kistler, Gerstner et al. 1997; Wilson 1999); however any simplification always comes with consequent reduction in biological fidelity and/or simulated neuron behavioural repertoire.

1.5.4 COUPLED EQUATIONS USING PARAMETERS WITH NO BIOPHYSICAL CORRELATES

If we assume that there is only one property of a neuron that needs to be modelled in order to capture everything about how one neuron influences others, and how neurons compute in general, then that property would be the membrane potential. In this case, the system of coupled equations can be greatly simplified while still exhibiting the rich set of spiking behaviors and sub-threshold characteristics of real neurons. The most successful model for this is the Izhikevich simple spiking neuron model (Izhikevich 2003), where success is measured in terms of both modelling efficiency and spiking behaviour. In fact, Izhikevich states that this is the simplest model possible which still exhibits all the required behaviours (Figure 1.9). It consists of just 2 equations and only 1 nonlinear term, so is computationally efficient – almost 100 times faster than the HH model to simulate. Studies have shown that the correspondence between a similar model of two variables (fitted to the data) and the HH models' spike timing predictions is very close – about 96% within 2 ms of each other (Brette and Gerstner 2005). The subthreshold dynamics of Izhikevich model neurons matches very closely those of HH (Boardman 2005).

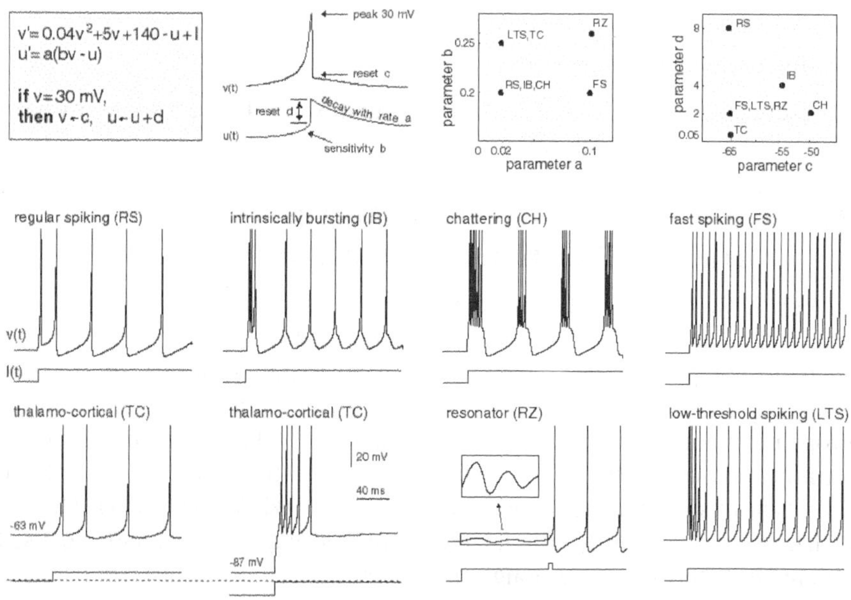

Figure 1.9 Top row – Izhikevich spiking neuron model. Bottom two rows – 8 of the 20 or so known neuron behaviours that can be emulated by the model.

The assumption that membrane potential is the only quantity needed for simulation of neural computation is arguable. A likely counter example is incorporation of synaptic plasticity into the model, which depends on concentrations of certain neurotransmitters and ions around the synapse (Urakubo and Watanabe 2002). However this doesn't rule out the possibility that these values could be reliably approximated by some simple function of the membrane potential or spike timings, keeping the model equally simple while maintaining functional fidelity. This is the approach taken in (Izhikevich 2005).

A range of other models exist that lie somewhere between Izhikevich and Hodgkin-Huxley, e.g., (FitzHugh 1961; Rose and Hindmarsh 1989), often by trying to simplify the HH computations while modelling more known physical neuron properties than just membrane potential. In general, it can be said that the more properties modelled, the more computation is required; it is therefore an implementation-dependent trade-off as to which model should be chosen. However, with its ability to exhibit all known biological neuron behaviours whilst remaining computationally economical, the Izhikevich model is the current standout choice.

1.6 INTEGRATE AND FIRE MODELS

The simplest integrate and fire (I&F) model – the leaky I&F neuron – was originally developed when dominant thinking stated that neuron function can be well-enough approximated by simply integrating input and then firing at a given threshold. Other properties like spike frequency adaptation, bursting, resonance, latency and variable thresholds were incorporated into models as needs arose and dominant thinking changed (see (Izhikevich 2004) for a review); however unfortunately no single I&F model displays all these characteristics.

The underlying assumption of all I&F models is that of neurons being integrators, which we now know is not always the case (e.g., resonance, bistability, inhibition-induced spiking). However they are computationally efficient, even more so than the Izhikevich model, and should be considered if the full range of spiking behaviour is not required.

1.6.1 SPIKING NEURAL NETWORKS – WHAT DO WE KNOW?

1.6.1.1 Delay coding and universal approximation

All Generation 2 neural networks (ie with continuous real-valued activation functions) can be emulated by SNNs, using delay coding (Maass 1997), also called latency or, somewhat confusingly, temporal or firing order coding. The process of delay coding is

- Determine the minimum and maximum activation levels of the Gen 2 neurons.
- Determine a suitable 'time-slice' length for the Gen 3 network that will equate to an iteration (one step) of the Gen 2 network. The Gen 3 network runs in continuous time made up of consecutive time-slices.
- Within each time-slice, fire each neuron at a time proportional to its Gen 2 activation level – e.g., neurons with maximum activation fire at the start

of the time-slice and neurons with minimum fire at the end; neurons with ¼ activation level fire ¾ of the way through the time-slice, and so on.

- Postsynaptic neurons decode the presynaptic activation level based on the firing delay within the time-slice.

Two obvious consequences of the fact that all ANNs can be emulated by SNNs are that

- Generation 2 ANNs are a subset of SNNs.
- SNNs are universal approximators (because Gen 2 ANNs are).

Another consequence is that

- Since any continuous function can be approximated with arbitrarily high reliability by an SNN with a single hidden layer (as can an MLP trained with back propagation), then with biologically realistic choices of spiking neuron parameters, any continuous function can be computed by an SNN within 20 ms (Maass 2001).

1.6.2 SPIKE TIMING DEPENDENT PLASTICITY (STDP)

STDP is a generalisation and refinement of Hebbian learning, stating that an increase (potentiation) in synaptic efficacy occurs if a presynaptic neuron fires immediately prior to the postsynaptic, and a decrease (depression) occurs if postsynaptic firing immediately precedes presynaptic. The effect was observed by Bi and Poo in hippocampal neurons in 1998 (Figure 1.10). The time course and magnitude of the effect vary between experimental studies, but it seems to last 10-50 ms either side of the postsynaptic spike and is maximal at or near the time of the spike, decaying exponentially to 0 by the end of the time course (Bi and Poo 1998; Kepecs, van Rossum et al. 2002; Urakubo and Watanabe 2002). It operates on excitatory synapses only and is less effective at potentiating already-strong synapses, although when synaptic depression occurs, the existing synaptic strength has less bearing on the result.

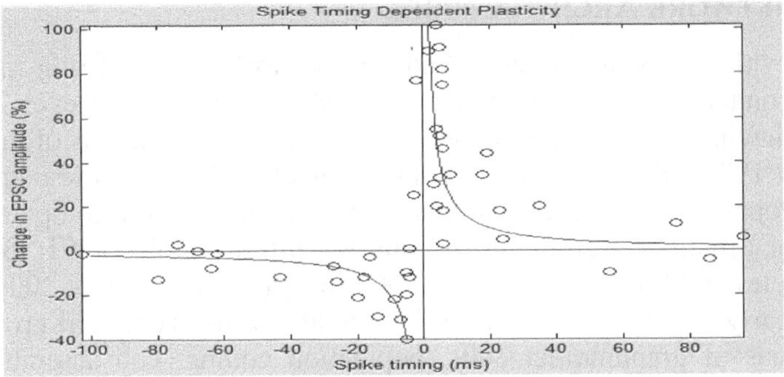

Figure 1.10 Change in Excitatory Post-Synaptic Current subsequent to different pre- and postsynaptic spike timings ($t_{post} - t_{pre}$). Exponential curves are shown for reference in red. Adapted from (Bi and Poo 1998).

Experimental data on the plasticity of inhibitory synapses on the other hand is scant (Swiercz, Cios et al. 2006), but often they are modelled more like Hebb's original rule whereby potentiation occurs if the two neurons fire closely together independent of the order of firing (Paugam-Moisy 2006). Others believe that inhibitory plasticity is computationally ill-advised and is not widespread in nervous systems (McBain, Freund et al. 1999).

STDP can be of itself lead to the stable development of network representations with no need for any form of weight or firing rate normalisation (Kempter, Gerstner et al. 1999; Song, Miller et al. 2000; van Rossum, Bi et al. 2000; Abbott and Gerstner 2004). However neurons are known to perform synaptic scaling in order to actively maintain a fixed long term average firing rate (Baddeley 1997); combined with standard Hebbian learning, this results in an implementation of Oja's rule, equivalent to principal components analysis or PCA (Oja 1982). STDP can be mathematically derived from several different starting constraints (Bohte and Mozer 2005). A multidisciplinary (molecular , biological and computational) and multiscale (of both temporal and spatial dimensions) review of STDP can be found in (Worgotter and Porr 2005).

Importantly, recent work has come up with the spiking neuron convergence conjecture (Legenstein, Naeger et al. 2005) which predicts that STDP can implement any input/output mapping that an SNN could ever potentially perform. They prove that the perceptron convergence theorem holds in the average case for STDP with Poisson input spike trains, then show through simulations that it holds in the test cases for more-realistic neurons and more-general input distributions. This endows STDP with universal learning capabilities.

1.6.3 NETWORK ARCHITECTURES

Neurons within nervous systems tend to have low average firing rates and be sparsely connected (Feldman and Ballard 1982; Brunel 2000; Reyes 2003), with synapses having a wide range of delays from one to several tens of milliseconds (Swadlow 1985). Rather than being a biological irrelevance, these appear to be crucial properties that affect how nervous systems perform computations. The concept of a temporal grouping of cells is central (Abeles 1991; Bienenstock 1995; Abeles 2002; Beggs and Plenz 2003; Ikegaya, Aaron et al. 2004), also see the very early paper (Rochester, Holland et al. 1956)! Temporal groups extend earlier ideas of grandmother cells, population coding, cell assemblies (Hebb 1949; Freeman 1991; Reilly 2001) and synfire chains (Abeles 1991) into the time domain since a group may not be a synchronously-firing collection of cells, but rather cells that fire in given order with known timings. Izhikevich calls this property 'polychrony' as against synchrony (Izhikevich 2005). Temporal groups come about due to sparse connectivity, spiking dynamics and particularly, random delays, and play a vital role in information processing by nervous systems, or in fact are the physical manifestation of the processing of information. So spiking neurons and their typical connectivity together support computation using SNNs, and it becomes arguably impossible, or at least unwise, to separate them.

The significance of the random delays for network function cannot be overstated. It's now well known that a sparsely connected network with random synaptic delays can easily have more states than the number of nodes (n) in the network (Izhikevich 2005) – remember a state in this terminology is not an attractor or stable state; rather it is a grouping of neurons that tend to fire in a given order over time. Contrast this to the Hopfield net (Hopfield 1982) where memories begin breaking down when just 0.13n for n = 100 (in general n/(4 log n) (McEliece, Posner et al. 1987)) are stored. In an SNN with well-chosen synaptic delays, there can actually be many more potential states than synapses in the entire network (Izhikevich 2005), which is an unprecedented memory capacity. Unfortunately, for systems of high-dimensional non-linear temporal interactions such as these, it is very difficult to conduct any rigorous mathematical analysis (Maass 2001), so no upper bound on memory capacity is known and no formal predictions of network behaviour can be made. Indeed, from a mathematical point of view, systems with temporal delays are infinite-dimensional. Despite the inability of off-the-shelf mathematics to provide analysis, it's clear that SNNs have unparalleled memory capacity.

Even randomly connected networks (or randomly connected satisfying certain constraints) with fixed (unitary) transmission delays exhibit potential to perform powerful computations, as long as there is a means to extract useful information from them. This is the basis of both Echo State Networks (Jaeger 2001) and Liquid State Machines (Maass, Natschlager et al. 2002), two closely related architectures that use random recurrent networks as their computational engines. Put very simply, their strategy is this:

- Pass the input into a highly re-entrant random network, which converts the input into a high dimensional dynamical representation. The weights and delays in the network are fixed, not trained, but in general there will be many more neurons than inputs.
- Simply train the output layer with least mean squares, linear regression or even a simple delta-rule variant.

These networks have been used to solve some quite difficult benchmark non-linear learning problems. The effectiveness of such a simple strategy is testament to the underlying computational power of recurrent networks, even random untrained ones. Put even more simply, the rationale behind their functioning is that, in any random network of useful yet still tractable size, the dynamics are rich enough that at least one neuron will have acquired close to any desired representation.

1.6.4 SPIKE CODING

The classic notion that spiking neurons encode information through their average spike rate over some time window (called a rate code) is obviously correct in some circumstances and obviously incorrect in others (Rieke, Steveninck et al. 1997). Sensory cells such as in the cochlear and the retina use a rate code (Izhikevich 2005), however response time in the visual cortex is known to be too fast to continue processing with this coding regime (Thorpe, Fize et al. 1996; Thorpe, Delorme et al. 2001) – each neuron in the visual processing hierarchy only has time to fire one or occasionally two spikes prior to recognition, so clearly the visual cortex cannot be using a rate code, but is instead somehow utilising the presence and/or the timing of spikes (called a temporal code, not to be confused with delay coding as described above, which is one form of temporal coding). Spike timing has been long-established as the information encoding principle used in the auditory system of bats for echolocation (Kuwabara and Suga 1993) and in the visual system of flies (Bialek, Rieke et al. 1991), for example.

There are a number of different coding strategies possible using spike times, shown below in increasing order of information encoding capacity (Thorpe, Delorme et al. 2001).

- **Count coding:** counts the total number of spikes of a neuron population in a given time – similar to rate coding except it entails one spike from each of many neurons instead of many spikes from one neuron; however the information capacity of rate coding is the same (very small).

- **Binary coding:** encodes a binary number, each digit represented by the presence (1) or absence (0) of a spike.
- **Rank order coding:** the order of firing of the neurons encodes the information.
- **Delay coding:** as described earlier.

Each of these coding strategies has very different information capacities summarized in the Table 1.2. In the table, n is the number of neurons under consideration, T is the time window over which each neuron can fire either 0 or 1 spikes, and p is the precision, which is the minimum interspike interval which can be discerned by postsynaptic neurons. In the row of the table labeled Example 1 it is assumed that n = 10 neurons, T = 10 ms and p = 1 ms. In Example 2, n = 1000 neurons while T and p remain unchanged, although this would overstate the information capacity of rank order coding since there are not n! discernable outcomes using this strategy in this case, as with 1 ms precision many spikes will appear to be simultaneous. In this case the number of discernable outcomes for rank order coding will actually be $\log_2({}^{1000}C_{100} \times {}^{900}C_{100} \times {}^{800}C_{100} \times \ldots \times {}^{100}C_{100}) = 3280$ bits (notice for Example 1 this degenerates to $\log_2 ({}^{10}C_1 \times {}^{9}C_1 \times {}^{8}C_1 \times \ldots \times {}^{1}C_1) = \log_2(10!)$). The information capacity of rank order coding is approaching that of delay coding (= 3320 bits) here because, with 1ms precision and 1000 spikes to squeeze into 10 ms, delays don't add a lot of extra information. Note that rank order coding also becomes indistinguishable from delay coding as the precision p approaches the mean interspike interval.

Table 1.2 Various descriptions and performance parameters

	Count coding	Binary coding	Rank order coding	Delay coding
Description	Counts the total number of spikes	A binary number	The order the neurons fire	The order and delays both encode information
Information capacity (bits)	$\log_2(n+1)$	n	$\log_2(n!)$	$n.\log_2(T/p)$
Example 1	3.6	10.0	21.8	33.2
Example 2	10.0	1000	?[3] (should be 3280)	3320

[3] The stated information capacity for rank order coding is not accurate when n > T/p since there cannot be n! discernible states in T milliseconds. See text for a more correct statement of information capacity in this case.

So which of the above temporal code strategies does the brain use when not using rate coding? For fast visual processing, rank order coding appears to suffice. For implementing STDP, delay coding would seem to be required in order to reliably control synaptic efficacy based on spike timings; however in nervous systems the distinction between order and delay coding may be blurred, since a delay of, for example, several hundred milliseconds between spikes from two neurons, does not change the order, yet is not likely to be interpreted as part of a single rank order code instance (i.e., even in order coding there are practical bounds on the delays).

Transmission delays in nervous systems display high variability even within connections that run in parallel, i.e., two connections that originate in one brain area, terminate in another brain area and follow very similar paths can have very different transmission delays. Conversely, some connection types such as thalamo-cortical can have very similar transmission delays irrespective of their length (Salami, Itami et al. 2003). It seems that nervous systems are able to exert purposeful control over these delays in situations where they may have computational significance. Some experimental data is shown in the Table 1.3 (Izhikevich 2005).

Table 1.3 Experimental Data based on Delay

Connection	Delay (ms)	Reference
Cat Layer 6 – LGN	1.0 – 44	(Ferster and Lindstrom 1983)
Rabbit Layer 6 – LGN	1.7 – 32	(Swadlow 1994)
Rabbit Cortico-cortical	1.0 – 35	(Swadlow 1985)
Rabbit Cortico-cortical	1.2 – 19	(Swadlow 1994)
Rabbit Cortico-(ipsi)cortical	2.2 – 32.5	(Swadlow 1994)
Rabbit Layer 5 – LGN	0.6 – 2.3	(Swadlow 1994)
Cat Cortico-collicular	0 – 3	(Ferster and Lindstrom 1983)
Mouse VB – Layer 4	2	(Salami, Itami et al. 2003)

Despite the high variability between connections in many brain areas, the delay in any given connection is consistently reproducible with sub-millisecond precision. The overarching theme here is that variable transmission delays are a fundamental property and a computational requirement of nervous systems, and hence should not be overlooked in SNN models.

1.6.5 COMPLEXITY

Interesting results have been obtained in analyses of the VC-dimension of spiking neurons (Maass and Schmitt 1997):

- The VC dimension of a threshold gate with n variable weights is $\Omega(n)$.

- The VC dimension of a spiking neuron with n variable delays is $\Omega(n.\log(n))$, even with fixed weights.

So the discriminatory power of variable delays is greater than that of variable weights. This implies that

- Networks with synaptic delays are potentially able to perform more powerful computations (Maass 1997; Maass 1997).
- Powerful learning algorithms could be formulated by adjusting synaptic delays in addition to or rather than synaptic weights.

Non-learnability results have been derived for SNNs that put the learning complexity of spiking neurons into the NP class of problems (Maass and Schmitt 1997). However due to the limitations of the mathematical tools that are used to conduct these rigorous analyses, simplifying assumptions must be made, and while these results mean it will be difficult to formally prove that learning algorithms work, it doesn't necessarily mean that they will be difficult to formulate.

1.6.6 SPIKING NEURAL NETWORKS – WHAT DON'T WE KNOW?

Despite the quite large body of knowledge expounded above, we still know very little about the dynamics, learning algorithms and computational abilities of SNNs. Obvious gaping holes in our knowledge include:

- **Broad mathematical analyses**: SNNs are high dimensional nonlinear dynamical systems. Consequently we generally cannot prove convergence for learning algorithms, and have little knowledge of upper bounds on memory capacities or of the dynamical behaviour of SNNs in differing circumstances. Given the richness of potential neuron behaviours, Izhikevich states in (Izhikevich 2004) "What happens when only tens (let alone billions) of such neurons are coupled together is beyond our comprehension". The (inadequate) alternative is to demonstrate these properties through exhaustive numerical simulation; however this proves nothing and it is difficult or impossible to generalise from these results.

- **Training of synaptic delays**: Given the computational power afforded by transmission delays in SNNs, there has been surprisingly little research published on mechanisms of training them. This may be partly due to the fact that there is dissent on whether or not the brain actually modifies delays as a facilitator of computational function (see (Eurich, Pawelzik et al. 1999) vs (Senn, Schneider et al. 2002) for example). However it has been shown that simple Hebbian-like learning rules can progressively modify synaptic delays so that spikes that arrive at different times within a given time window can ultimately be synchronised at the postsynaptic neuron, and produce stable representations of temporal input (Huning, Glunder et al. 1998; Eurich, Pawelzik et al. 1999). These rules function by adding to a delay if the presynaptic neuron fires some time before the postsynaptic, and subtracting from a delay if the presynaptic neuron fires some time later. Biological mechanisms for controlling the delay can include changing the thicknesses of axons and dendrites or the extent of myelination (Fields 2005). An alternative to actively adapting transmission delays is to simply select appropriate delays from an initial over-abundance of random delays; delays that by chance closely match the input train are strengthened by normal Hebbian learning, while the remaining ineffective delays are depressed by the same mechanism and may ultimately vanish. It is well known that the juvenile brain contains many more synapses than the adult, but whether selection of delays is one reason for this reduction over time, and whether delay training is also undertaken by the brain, are still open questions.

- **How does the brain compute?** Transmission delays, spike coding, temporal grouping, nested oscillations, phase precession – irrespective of what we do or don't know about spiking neurons, some of the most fundamental attributes of the brain are still almost complete mysteries to us. Although SNNs are the next step towards a full understanding of nervous system computation and can help us to model many of these till-now neglected properties, they are not likely to be the ultimate level of detail required to model all of brain function, and we may yet need Generation 4 and beyond neural networks for this task.

1.6.7 WHO IS USING SPIKING NEURONS?

Work with spiking neurons and SNNs has been going on at the periphery of neural network research for many years. This includes:

- A considerable amount of theoretical work by mathematicians and physicists, some of which has been discussed above; see (Rieke, Steveninck et al. 1997) for more information.
- Obviously all neuroscientists, who are most interested in the physical mechanisms of spiking, STDP etc, hence use spiking neuron models with high biological realism.
- Brain region modelling – deep but not broad models of targeted brain regions operating in specific modalities e.g., hippocampus in a specific navigation task (Hasselmo, Bodelon et. al., 2002).
- Cognitive modelling – understandably, there is a disconnect between higher level brain models and models of individual spikes (solving this completely is arguably solving AI!)

- Some applications – currently dominated by auditory (mostly speech) processing (Hopfield and Brody 2000; Hopfield and Brody 2001; Loiselle, Rouat et al. 2005; Verstraeten, Schrauwen et al. 2005) and visual processing (Perrinet and Samuelides 2002; Azhar, Iftekharuddin et al. 2005; Kornprobst, Vieille et al. 2005); SpikeNet Technology is a commercialised vision package (Thorpe and Gautrais 1997; Thorpe, Guyonneau et al. 2004), also see http://www.spikenet-technology.com. Robotics with SNNs is just beginning to heat up (Di Paolo 2002; Nielsen and Lund 2003; Roggen, Hofmann et al. 2003; Floreano, Epars et al. 2005; Floreano, Zufferey et al. 2005), although most current robotics implementations depend on evolutionary algorithms to create the SNNs.

1.6.8 CONCLUSION

With the knowledge we are currently obtaining of the fundamental importance of spike timings and oscillations to neural processing, 2nd generation ANNs can no longer provide a viable basis for neural modelling. Spiking Neural Networks present many new challenges but also afford many new opportunities for breaking entirely new ground in artificial intelligence research.

1.7 APPLICATIONS OF ANN

Let us suppose we wish to train an ANN to recognize handwriting. For our purposes we wish to train it to recognize the letters H and C. More exactly for this simple application we would be happy for it to be able to identify an H or a C. Handwriting recognition is a problem that neural networks have been able to tackle very well. The issues that arise here are common to all neural network classification problems.

The first activity is to represent the letters for input to the ANN. Typically we draw a grid over the letter and represent the inputs as zero or one depending on whether the hand-written letter goes through the cell in the grid. Figure 1.11 gives an example for an H and a C.

The inputs will be strings of zeros and ones. The string is achieved by going from the top left-hand corner across to the right and then left to right a row at a time, where a 1 indicates that the letter goes through that square of the grid. In the example shown, the inputs will be 110111011 for H and 110100111 for C. We would only need one output for this particular problem. We could perhaps give it a value of 1 for an H and 0 for a C. So there are nine inputs and one

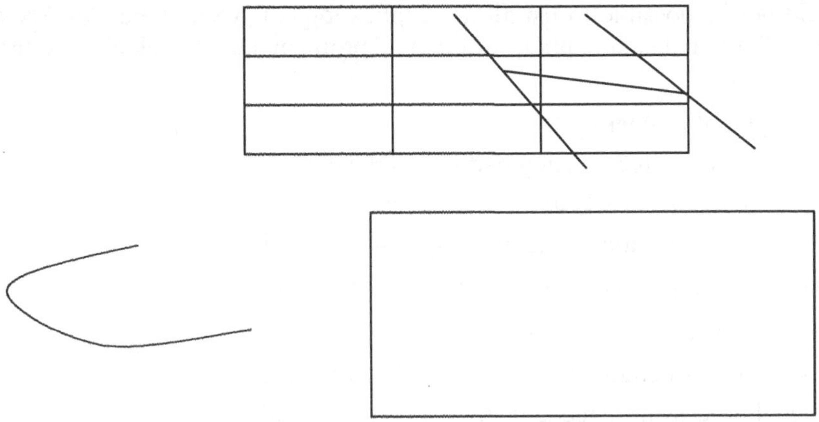

Figure 1.11 Letter Recognition - an H and a C

output. For most applications we only require one hidden layer. As to how many neurons there should be in the hidden layer this is still a case of 'trial and error'. Typically you experiment from, say, about 7 down to 2. Suppose we have 4 in the hidden layer. This is called a 9-4-1 network. This network has, therefore, 40 [i.e. $(9 \times 4) + (4 \times 1)$] weights that have to be learned. For a neural network to have good generalisation capabilities (to be able to classify inputs it has not seen before) the literature reports that you need approximately 5-10 times as many training pairs as weights. So in this example we would need about 200 letters with approximately 100 Hs and 100 Cs. As you can see, if you dropped to 2 neurons in the hidden layer you would need only 100 data sets. The questions that need answering then are:

1. How do we represent the data for input?
2. How many neurons should be in the output layer and, in the case of classification problems, what values will those neurons take?
3. How many neurons should be in the hidden layer?
4. What activation function should we use for the neuron?
5. What is an acceptable error rate?

This section has introduced you to the most common paradigm in neural network applications – the feed-forward, multilayered perceptron with the backpropagation algorithm. This approach is just one example of a supervised algorithm.

SUMMARY OF ANN

Artificial Neural Networks – What are They good for?

It would not be possible to list all the applications of ANNs here. An ANN approach will often be an option where the problem being tackled has the following features.

- The type of problem:
    - is one of recognising patterns in the data.
    - requires classification of data into, for example, classes.
    - is one of monitoring of equipment in real time.
- There is a large amount of data that may be 'noisy'.

Some example applications are:

- Monitoring of engine condition in a fleet of vehicles.
- Signal processing – recognising patterns in signals.
- Face recognition. Neural networks are particularly good at recognising shapes, e.g. fingerprints, signatures, tanks on an horizon.
- Process control - using ANNs to monitor equipment.
- Forecasting corporate bankruptcy based on financial indicators.
- Credit scoring to assess credit worthiness when considering giving a loan.
- We know how neural networks work and the types of applications for which they are suitable.

Artificial Neural Networks - What are the Drawbacks?

- They require large amounts of historical data that accurately reflects the make-up of the population under consideration.
- They are 'black box'. Unlike expert systems, they are incapable of explaining why they make a particular decision. This is a major problem when trying to 'sell' neural network technology to management. The only way to test the efficacy of an ANN solution is to test the trained network with many examples that it has not seen before.
- There are a large number of parameters that the ANN developer has to make decisions about. For example he/she has to decide on the learning rate, activation functions, the network structure or topology, how to represent the problem, etc.

In summary, ANNs are a powerful, practical solution to many problems faced by industry and commerce and should be considered as one of the tools in the armoury of the professional trying to find solutions to difficult problems.

REFERENCES

1. Aarts, E.H., F.M.J. de Bont, E. H. A. Haberrs, P.J. M. Laarhoven. (1986), "A Parallel Statistical Cooling Algorithm", *Lecture Note in Computer Science* 210, pp 87

2. Aarts E., anf J. Korst, (1989), *Simulated Annealing and Boltzmann Machines: A Stochastic Approach to Combinatorial Optimization and Neural Computing,* New York: Wiley

3. Abe S. (1989) "Theories of the Hopfield Neural Networks", *Proc. International Joint Conference on Neural Networks* (IJCNN'89), Washington D.C., Vol. I, pp 557- 564, June

4. Abe S. (1991) "Determining Weights of the Hopfield Neural Networks", *Proc.International Conference on Artificial Neural Networks* (ICANN'91) Helsinki, pp 1507-1510, June

5. Abe S. "Global Convergence and Supression of Spurious States of the Hopfield Neural Networks ", *Trans. IEEE Circuits & Systems,*

6. Abraham, R.H., and C.D. Shaw, (1992), *Dynamics of the Geometry of Behavior*, Reading, MA, Addison-Wesley.

7. Aiyer, S.V.B., M. Niranjan, F. Fallside, (1990), "On the Optimization Properties of the Hopfield Model", *Proc. International Conference on Neural Networks* (ICNN'90), pp 245-249

8. Aiyer S.V.B., N. Niranjan and F. Fallside, (1990), "A Theoretical Investigation into the Performance of the Hopfield Model.", *IEEE Transactions on Neural Networks* 15, 15, 204-215

9. Akiyema et al, (1991) "The Gaussian Machine: A stochastic Neural Network for Solving Assignment Problems", *Journal of Neural Network Computing,* Winter, pp 43-51

10. Allwright, J.R.A. and D.B. Carpenter, (1989), "A Distributed Implementation of Simulated Annealing for the Traveling Salesman Problem", *Parallel Computing* 10, pp 335, North Holland

CHAPTER 2

FUNDAMENTAL OF NEURAL NETWORKS

2.1 INTRODUCTION

For many decades, it has been a goal of science and engineering to develop intelligent machines with a large number of simple elements. References to this subject can be found in the scientific literature of the 19th century. During the 1940s, researchers desiring to duplicate the function of the human brain, have developed simple hardware (and later software) models of biological neurons and their interaction systems. McCulloch and Pitts [1] published the first systematic study of the artificial neural network. Four years later, the same authors explored network paradigms for pattern recognition using a single layer perceptron [2]. In the 1950s and 1960s, a group of researchers combined these biological and psychological insights to produce the first artificial neural network (ANN) [3,4]. Initially implemented as electronic circuits, they were later converted into a more flexible medium of computer simulation. However, researchers such as Minsky and Papert [5] later challenged these works. They strongly believed that intelligence systems are essentially symbol processing of the kind readily modeled on the Von Neumann computer. For a variety of reasons, the symbolic–processing approach became the dominant method. Moreover, the perceptron as proposed by Rosenblatt turned out to be more limited than first expected [4]. Although further investigations in ANN continued during the 1970s by several pioneer researchers such as Grossberg, Kohonen, Widrow, and others, their works received relatively less attention. The primary factors for the recent resurgence of interest in the area of neural networks are the extension of Rosenblatt, Widrow and Hoff's works dealing with learning in a complex, multi-layer network, Hopfield mathematical foundation for understanding the dynamics of an important class of networks, as well as much faster computers than those of 50s and 60s.

The interest in neural networks comes from the networks' ability to mimic human brain as well as its ability to learn and respond. As a result, neural networks have been used in a large number of applications and have proven to be effective in performing complex functions in a variety of fields. These include pattern recognition, classification, vision, control systems, and prediction [6], [7]. Adaptation or learning is a major focus of neural net research that provides a degree of robustness to the NN model. In predictive modeling, the goal is to map a set of input patterns onto a set of output patterns. NN accomplishes this task by learning from a series of input/output data sets presented to the network. The trained network is then used to apply what it has learned to approximate or predict the corresponding output [8].

This chapter is organized as follows. In section 2.2, various elements of an artificial neural network are described. The Adaptive Linear Element (ADALINE) and single layer perceptron are discussed in section 2.3 and 2.4 respectively. The multi-layer perceptron is presented in section 2.5. Section 2.6 discusses multi-layer perceptron and section 2.7 concludes this chapter.

2.2 BASIC STRUCTURE OF A NEURON

2.2.1 MODEL OF BIOLOGICAL NEURONS

In general, the human nervous system is a very complex neural network. The brain is the central element of the human nervous system, consisting of near 1×10^{10} biological neurons that are connected to each other through sub-networks. Each neuron in the brain is composed of a body, one axon and multitude of dendrites. The neuron model shown in Figure 2.1 serves as the basis for the artificial neuron. The dendrites receive signals from other neurons. The axon can be considered as a long tube, which divides into branches terminating in little endbulbs. The small gap between an endbulb and a dendrite is called a synapse. The axon of a single neuron forms synaptic connections with many other neurons. Depending upon the type of neuron, the number of synapses connections from other neurons may range from a few hundreds to 10^4.

The cell body of a neuron sums the incoming signals from dendrites as well as the signals from numerous synapses on its surface. A particular neuron will send an impulse to its axon if sufficient input signals are received to stimulate the neuron to its threshold level. However, if the inputs do not reach the required threshold, the input will quickly decay and will not generate any action. The biological neuron model is the foundation of an artificial neuron as will be described in detail in the next section.

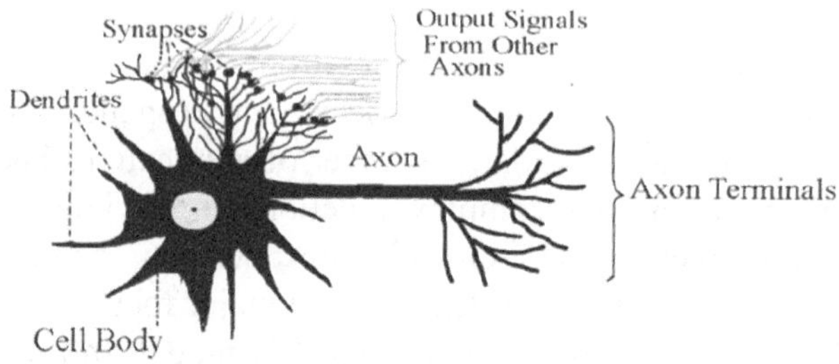

Figure 2.1 A Biological Neuron.

2.2.2 ELEMENTS OF NEURAL NETWORKS

An artificial neuron as shown in Figure 2.2, is the basic element of a neural network. It consists of three basic components that include weights, thresholds, and a single activation function.

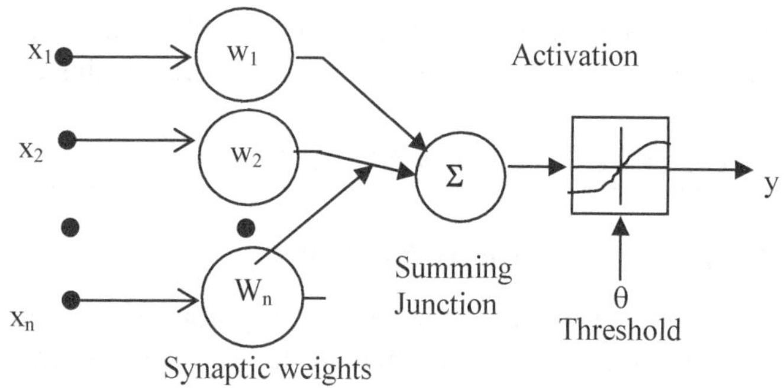

Figure 2.2 Basic Elements of an Artificial Neuron.

2.2.2.1 Weighting Factors

The values W_1 ,W_2 ,W_3 ,.....,W_n are weight factors associated with each node to determine the strength of input row vector $X = [x_1\ x_2\ x_3.....,x_n]^T$. Each input is multiplied by the associated weight of the neuron connection X^TW. Depending upon the activation function, if the weight is positive, X^TW commonly excites the node output; whereas, for negative weights, X^TW tends to inhibit the node output.

2.2.2.2 Threshold

The node's internal threshold θ is the magnitude offset that affects the activation of the node output y as follows:

$$y = \sum_{i=1}^{n}(X_iW_i) - \theta_k \qquad \text{..... (2.1)}$$

where i, n, θ_k, X_i, W_i.......

2.2.2.3 Activation Function

In this subsection, five of the most common activation functions are presented. An activation function performs a mathematical operation on the signal output. More sophisticated activation functions can also be utilized depending upon the type of problem to be solved by the network. All the activation functions as described herein are also supported by MATLAB package.

Linear Function

As is known, a linear function satisfies the superposition concept. The function is shown in Figure 2.3.

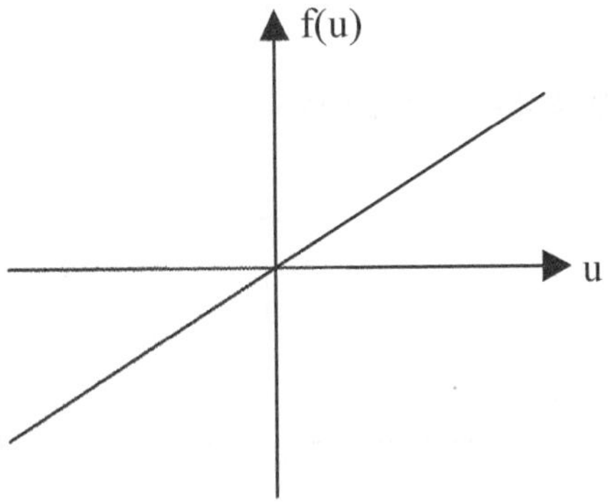

Figure 2.3 Linear Activation Function.

The mathematical equation for the above linear function can be written as

$$y = f(u) = \alpha . u \qquad .....(2.2)$$

where α is the slope of the linear function 2.2. If the slope α is 1, then the linear activation function is called the identity function. The output (y) of identity function is equal to input function (u). Although this function might appear to be a trivial case, nevertheless it is very useful in some cases such as the last stage of a multilayer neural network.

Threshold Function

A threshold (hard-limiter) activation function is either a *binary* type or a *bipolar* type as shown in Figures 2.4 and 2.5, respectively. The output of a *binary* threshold function can be written as:

$$y = f(u) = \begin{cases} 0 & if \quad u < 0 \\ 1 & if \quad u \geq 0 \end{cases} \qquad .....(2.3)$$

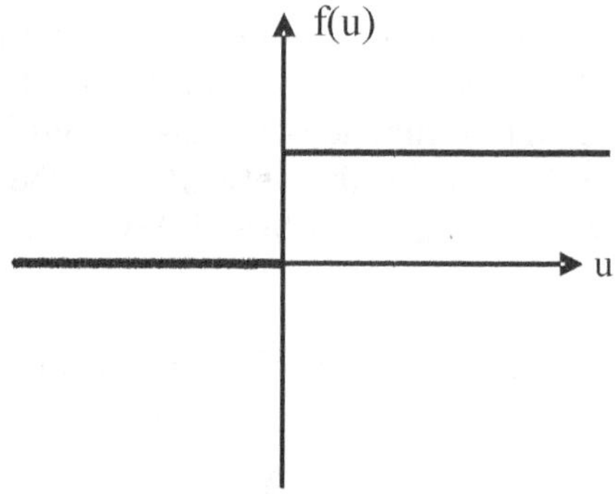

Figure 2.4 Binary Threshold Activation Function.

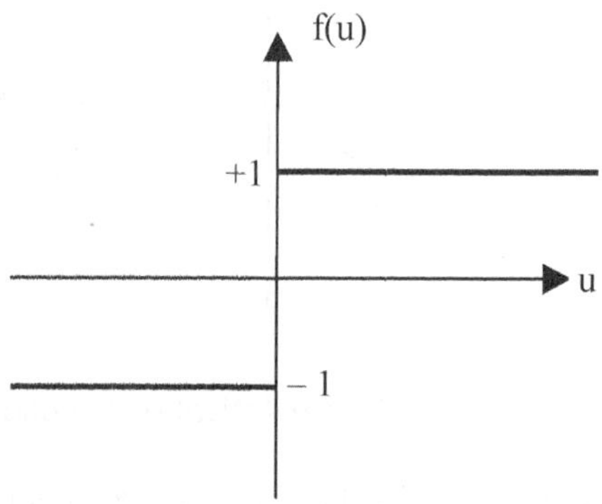

Figure 2.5 Bipolar Threshold Activation Function.

The neuron with the hard limiter activation function is referred to as the McCulloch-Pitts model.

Piecewise Linear Function

This type of activation function is also referred to as saturating linear function and can have either a binary or bipolar range for the saturation limits of the output. The mathematical model for a symmetric saturation function (Figure 2.6) is described as follows:

$$y = f(u) = \begin{cases} -1 & if \quad u < -1 \\ u & if \quad -1 \geq u \geq 1 \\ 1 & if \quad u \geq 1 \end{cases} \qquad \text{.....(2.4)}$$

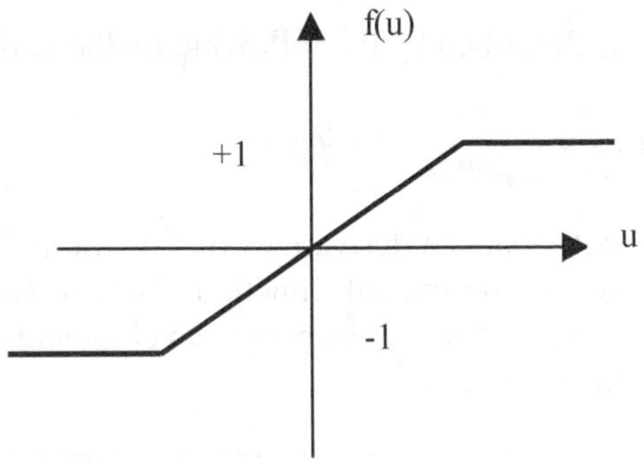

Figure 2.6 Piecewise Linear Activation Function.

Sigmoidal (S shaped) function

This nonlinear function is the most common type of the activation used to construct the neural networks. It is mathematically well behaved, differentiable and strictly increasing function. A sigmoidal transfer function can be written in the following form:

$$f(x) = \frac{1}{1+e^{-\alpha x}}, \quad 0 \le f(x) \le 1 \qquad .....(2.5)$$

where α is the shape parameter of the sigmoid function. By varying this parameter, different shapes of the function can be obtained as illustrated in Figure 2.7. This function is continuous and differentiable.

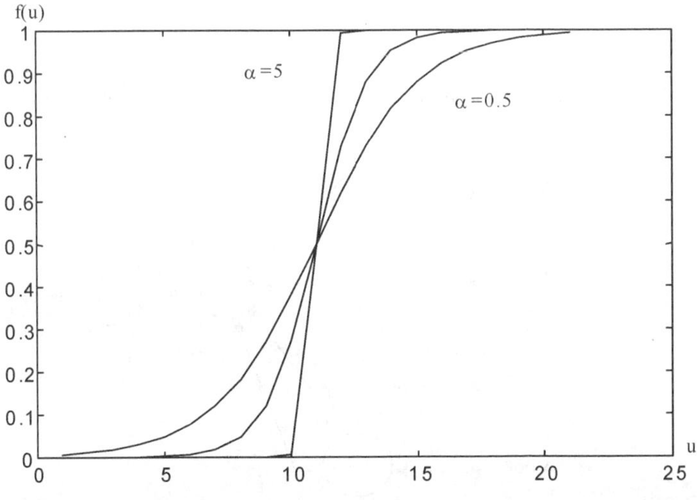

Figure 2.7 A Sigmoid Activation Function.

Tangent hyperbolic function

This transfer function is described by the following mathematical form:

$$f(x) = \frac{e^{\alpha x} - e^{-\alpha x}}{e^{\alpha x} + e^{-\alpha x}} \quad -1 \le f(x) \le 1 \qquad .....(2.6)$$

It is interesting to note that the derivatives of Equations 2.5 and 2.6 can be expressed in terms of the individual function itself (please see problems appendix). This is important for the learning development rules to train the networks as shown in the next chapter.

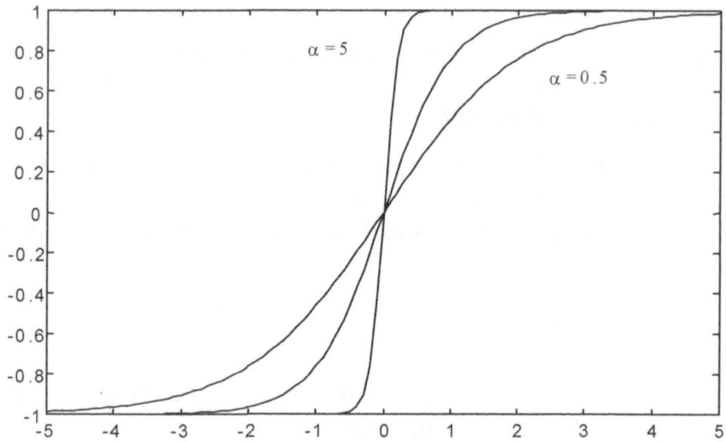

Figure 2.8 A Tangent Hyperbolic Activation Function.

Example 2.1

Consider the following network consists of four inputs with the weights as shown

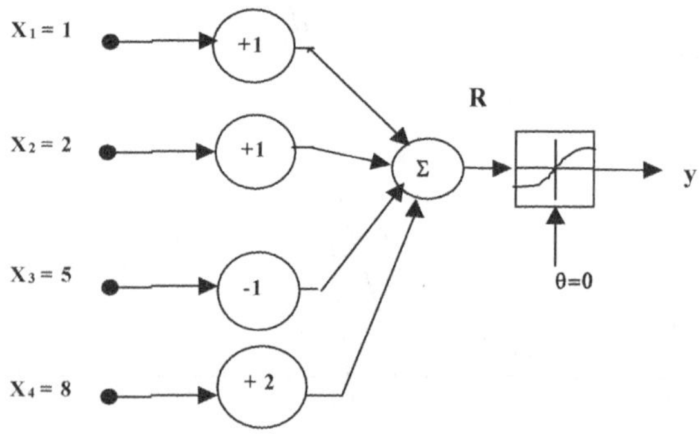

Figure 2.9 Neuron Structure of Example 2.1.

The output R of the network, prior to the activation function stage, is calculated as follows:

$$R = W^T.X = \begin{bmatrix} 1 & 1 & -1 & 2 \end{bmatrix} . \begin{bmatrix} 1 \\ 2 \\ 5 \\ 8 \end{bmatrix} = 14 \qquad .....(2.7)$$

With a binary activation function, and a sigmoid function, the outputs of the neuron are respectively as follow:

y(*Threshold*) = 1;

y(*Sigmoid*) = 1.5×2^{-8}

2.3 ADALINE

An ADAptive LINear Element (ADALINE) consists of a single neuron of the McCulloch-Pitts type, where its weights are determined by the normalized least mean square (LMS) training law. The LMS learning algorithm was originally proposed by Widrow and Hoff [6]. This learning rule is also referred to as delta rule. It is a well-established supervised training method that has been used over a wide range of diverse applications [7]- [11]. Curve fitting approximations can also be used for training a neural network [10]. The learning objective of curve fitting is to find a surface that best fits to the training data. In the next chapter the implementation of LMS algorithms for backpropagation, and curve fitting algorithms for radial basis function network, will be described in detail.

The architecture of a simple ADALINE is shown in Figure 2.10. It is observed that the basic structure of an ADALINE is similar to a linear neuron (Figure 1.2) with the activation function f(.) to be a linear one with an extra feedback loop. Since ADALINE is a linear device, any combination of these units can be accomplished with the use of a single unit.

During the training phase of ADALINE, the input vector $X \in R^n$: $X = \begin{bmatrix} x_1 & x_2 & x_3 & \cdots & x_n \end{bmatrix}^T$ as well as desired output are presented to the network. The weights are adaptively adjusted based on delta rule. After the ADALINE is trained, an input vector presented to the network with fixed weights will result in a scalar output. Therefore, the network performs a mapping of an *n* dimensional mapping to a scalar value. The activation function is not used during the training phase. Once the weights are properly adjusted, the response of the trained unit can be tested by applying various inputs, which are not in the training set. If the network produces consistent responses to a high degree with the test inputs, it is said that the network could *generalize*.

Therefore, the process of training and generalization are two important attributes of the network.

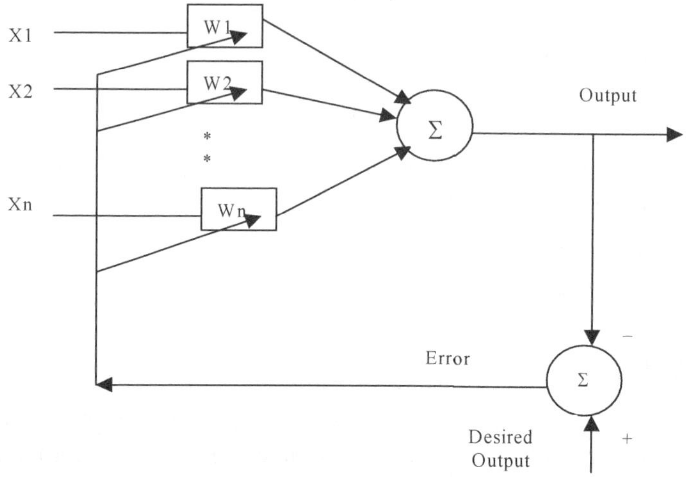

Figure 2.10 ADALINE.

In practice, an ADALINE is usually used to make binary decisions. Therefore, the output is sent through a binary threshold as shown in Figure 2.4. Realizations of several logic gates such as AND, NOT and OR are common applications of ADALINE. Only those logic functions that are linearly separable can be realized by the ADALINE, as is explained in the next section.

2.4 LINEAR SEPARABLE PATTERNS

For a single ADALINE to function properly as a classifier, the input pattern must be linearly separable. This implies that the patterns to be classified must be sufficiently apart from each other to ensure the decision surface consists of a single hyperplane such as a single straight line in two-dimensional space. This concept is illustrated in Figure 2.11 for a two-dimensional pattern.

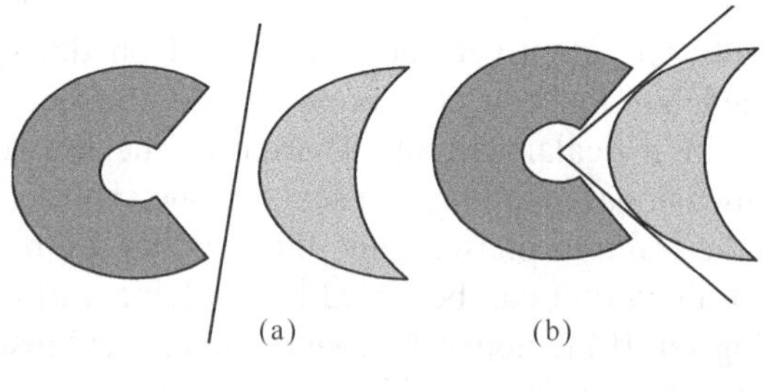

Figure 2.11 (a) A Pair of Linearly Separable (b), and Non-Linearly Separable Patterns.

A classic example of a mapping that is not separable is XOR (the exclusive or) gate function. Table 2.1 shows the input-output pattern of this problem. Figure 2.12 shows the locations of the symbolic outputs of XOR function corresponding to four input patterns in X_1-X_2 plane. There is no way to draw a single straight line so that the circles are on one side of the line and the triangular sign on the other side. Therefore, an ADALINE cannot realize this function.

Table 2.1 Inputs/Outputs Relationship for XOR.

X_1	X_2	*Output*
0	0	0
0	1	1
1	0	1
1	1	0

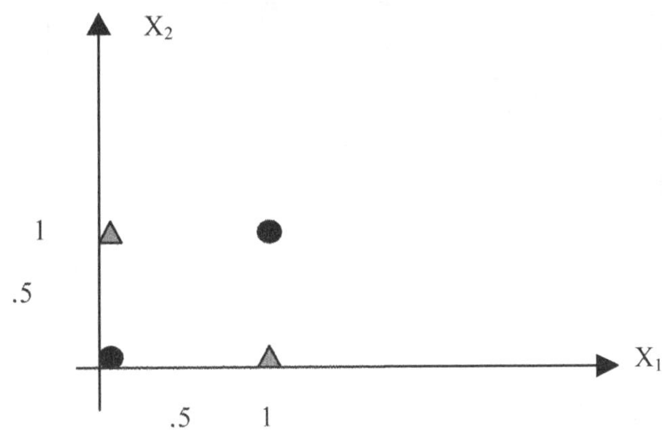

Figure 2.12 The Output of XOR in X1-X2 Plane.

One approach to solve this nonlinear separation problem is to use MADALINE (Multiple ADALINE) networks. The basic structure of a MADALINE network consists of combining several ADALINE with their correspondence activation functions into a single forward structure. When suitable weights are chosen, the network is capable of implementing complicated and nonlinear separable mapping such as XOR gate problems. We will address this issue later in this chapter.

2.5 SINGLE LAYER PERCEPTRON

2.5.1 GENERAL ARCHITECTURE

The original idea of the perceptron was developed by Rosenblatt in the late 1950s along with a convergence procedure to adjust the weights. In Rosenblatt's perceptron, the inputs were binary and no bias was included. It was based on the McCulloch-Pitts model of the neuron with the hard limitation activation function. The single layer perceptron as shown in Figure 2.13 is very similar to ADALINE except for the addition of an activation function.

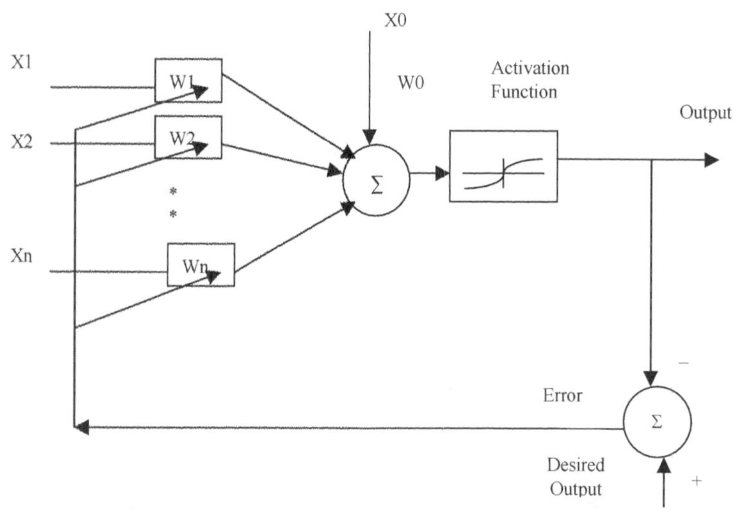

Figure 2.13 A Perceptron with a Sigmoid Activation Function.

Connection weights and threshold in a perceptron can be fixed or adapted using a number of different algorithms. Here the original perceptron convergence procedure as developed by Minsky and Papert[5] is described. First, connection weights W_1, W_2,...,W_n and the threshold value W_0 are initialized to small non-zero values. Then, a new input set with N values received through sensory units (measurement devices) and the input is computed. Connection weights are only adapted when an error occurs. This procedure is repeated until the classification of all inputs is completed.

2.5.2 LINEAR CLASSIFICATION

For clarification of the above concept, consider two input patterns classes C1 and C2. The weight adaptation at the kth training phase can be formulated as follow:

1. If k member of the training vector x(k) is correctly classified, no correction action is needed for the weight vector. Since the activation function is selected as a hard limiter, the following conditions will be valid:

   $$W(k+1) = W(k) \text{ if } output>0 \text{ and } x(k) \in C1, \text{ and}$$

   $$W(k+1) = W(k) \text{ if } output<0 \text{ and } x(k) \in C2.$$

2. Otherwise, the weight should be updated in accordance with the following rule:

   $$W(k+1) = W(k) + \eta\, x(k) \text{ if } output \geq 0 \text{ and } x(k) \varepsilon\, C1$$

   $$W(k+1) = W(k) - \eta\, x(k) \text{ if } output \leq 0 \text{ and } x(k) \varepsilon\, C2$$

   Where η is the learning rate parameter, which should be selected between 0 and 1.

Example 2.2

Let us consider pattern classes C1 and C2, where C1: {(0,2), (0,1)} and C2: {(1,0), (1,1)}. The objective is to obtain a decision surface based on perceptron learning. The 2-D graph for the above data is shown in Figure 2.14

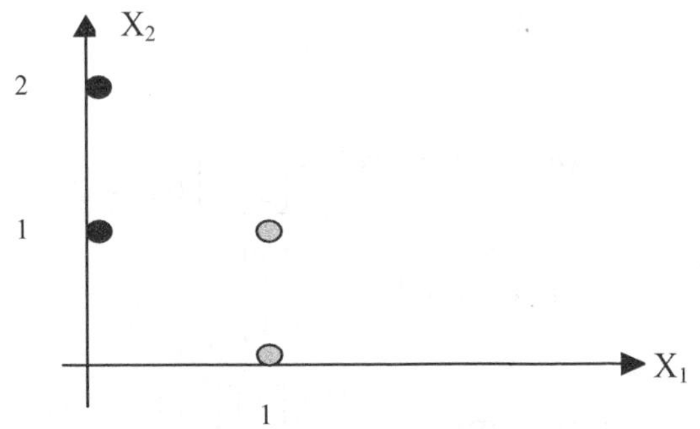

Figure 2.14 2-D Plot of Input Data Sets for Example 2.2.

Since, the input vectors consist of two elements, the perceptron structure is simply as follows:

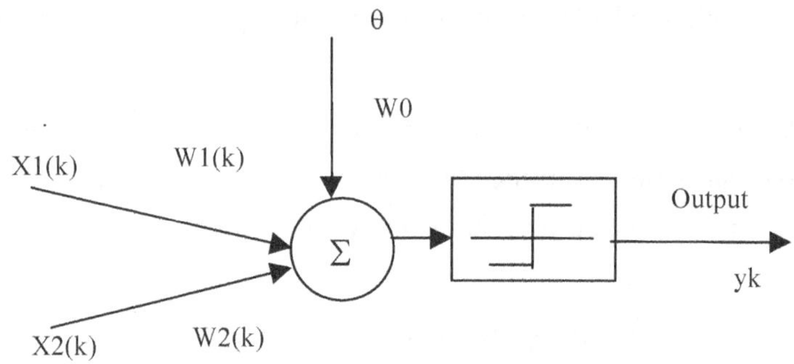

Figure 2.15 Perceptron Structure for Example 2.2.

For simplicity, let us assume η = 1 and initial weight vector W(1) = [0 0]. The iteration weights are as follow:

Iteration 1: $W^T(1).x(1) = \begin{bmatrix} 0 & 0 \end{bmatrix} \begin{bmatrix} 0 \\ 2 \end{bmatrix} = 0ax$

Weight Update: $W(2) = W(1) + x(1) = \begin{bmatrix} 0 \\ 0 \end{bmatrix} + \begin{bmatrix} 0 \\ 2 \end{bmatrix} = \begin{bmatrix} 0 \\ 2 \end{bmatrix}$

Iteration 2: $W^T(2).x(2) = \begin{bmatrix} 0 & 2 \end{bmatrix} \begin{bmatrix} 0 \\ 1 \end{bmatrix} = 2 > 0$

Weight Update: $W(3) = W(2)$

Iteration 3: $W^T(3).x(3) = \begin{bmatrix} 0 & 2 \end{bmatrix} \begin{bmatrix} 1 \\ 0 \end{bmatrix} = 0$

Weight Update: $W(4) = W(3) - x(3) = \begin{bmatrix} 0 \\ 2 \end{bmatrix} - \begin{bmatrix} 1 \\ 0 \end{bmatrix} = \begin{bmatrix} -1 \\ 2 \end{bmatrix}$

Iteration 4: $W^T(4).x(4) = \begin{bmatrix} -1 & 2 \end{bmatrix} \begin{bmatrix} 1 \\ 1 \end{bmatrix} = 1$

Weight Update: $W(5) = W(4) - x(4) = \begin{bmatrix} -1 \\ 2 \end{bmatrix} - \begin{bmatrix} 1 \\ 1 \end{bmatrix} = \begin{bmatrix} -2 \\ 1 \end{bmatrix}$

Now if we continue the procedure, the perceptron classifies the two classes correctly at each instance. For example for the fifth and sixth iterations:

Iteration 5: $W^T(5).x(5) = [-2 \quad 1]\begin{bmatrix}0\\2\end{bmatrix} = 2 > 0$:*Correct Classification*

Iteration 6: $W^T(6).x(6) = [-2 \quad 1]\begin{bmatrix}0\\1\end{bmatrix} = 1 > 0$:*Correct Classification*

In a similar fashion for the seventh and eighth iterations, the classification results are indeed correct.

Iteration 7: $W^T(7).x(7) = [-2 \quad 1]\begin{bmatrix}1\\0\end{bmatrix} = -2 < 0$:*Correct Classification*

Iteration 8: $W^T(8).x(8) = [-2 \quad 1]\begin{bmatrix}1\\1\end{bmatrix} = -1 < 0$:*Correct Classification*

Therefore, the algorithm converges and the decision surface for the above perceptron is as follows:

$$d(x) = -2X_1 + X_2 = 0 \qquad .....(2.8)$$

Now, let us consider the input data {1,2}, which is not in the training set. If we calculate the output:

$$Y = W^T.X = [-2 \quad 1]\begin{bmatrix}2\\1\end{bmatrix} = -3 < 0 \qquad .....(2.9)$$

The output Y belongs to the class C2 as is expected.

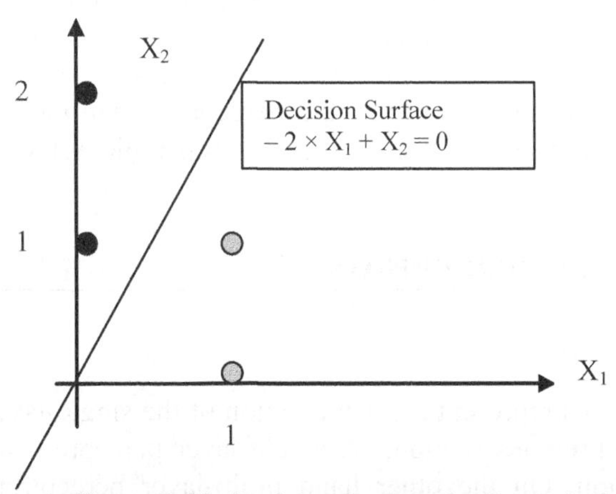

Figure 2.16 Decision Surface for Example 2.2.

2.5.3 PERCEPTRON ALGORITHM

The perceptron learning algorithm (Delta rule) can be summarized as follows:

Step 1: Initialize the weights W_1, $W_2 \ldots W_n$ and threshold θ to small random values.

Step 2: Present new input X1, X2,..Xn and desired output d_k.

Step 3: Calculate the actual output based on the following formula:

$$y_k = f_h\left(\sum_{i=1}^{n}(X_i Wi) - \theta_k\right) \qquad \text{.....(2.10)}$$

Step 4: Adapt the weights according to the following equation:

$$W_i(new) = W_i \ (old) + \eta(d_k - y_k)x_i, 0 \le i \le N \qquad \text{.....(2.11)}$$

Where η is a positive gain fraction less than 1 and d_k is the desired output. Note that the weights remain the same if the network makes the correct decision.

Step 5: Repeat the procedures in steps 2–4 until the classification task is completed.

Similar to ADALINE, if the presented inputs pattern is linearly separable, then the above perceptron algorithm converges and positions the decision hyperplane between two separate classes. On the other hand, if the inputs are not separable and their distribution overlaps, then the decision boundary may oscillate continuously. A modification to the perceptron convergence procedure is the utilization of Least Mean Square (LMS) in this case. The algorithm that forms the LMS solution is also called the Widrow-Hoff. The LMS algorithm is similar to the procedure above except a threshold logic nonlinearity, replaces the hard limited non-linearity. Weights are thus corrected on every trail by an amount that depends on the difference between the desired and actual values. Unlike the learning in the ADALINE, the perceptron learning rule has been shown to be capable of separating any linear separable set of the training patterns.

2.6 MULTI-LAYER PERCEPTRON

2.6.1 GENERAL ARCHITECTURE

Multi-layer perceptrons represent a generalization of the single-layer perceptron as described in the previous section. A single layer perceptron forms a half–plane decision region. On the other hand multi-layer perceptrons can form arbitrarily complex decision regions and can separate various input patterns.

The capability of multi-layer perceptron stems from the non-linearities used within the nodes. If the nodes were linear elements, then a single-layer network with appropriate weight could be used instead of two- or three-layer perceptrons. Figure 2.17 shows a typical multi-layer perceptron neural network structure. As observed it consists of the following layers:

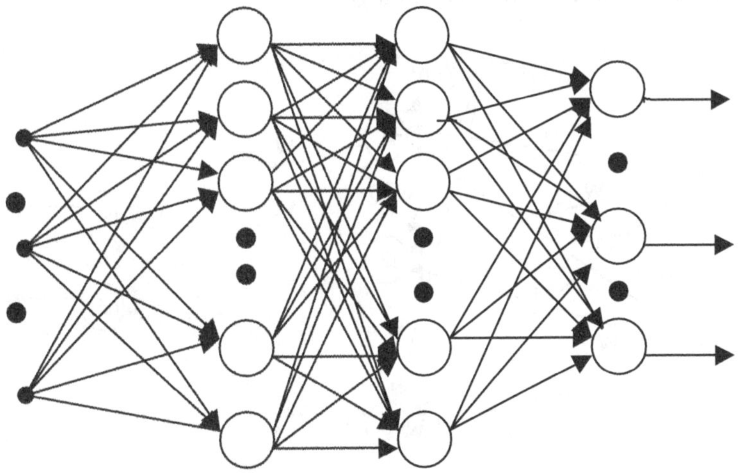

Figure 2.17 Multi-layer Perceptron.

***Input Layer*:** A layer of neurons that receives information from external sources, and passes this information to the network for processing. These may be either sensory inputs or signals from other systems outside the one being modeled.

***Hidden Layer*:** A layer of neurons that receives information from the input layer and processes them in a hidden way. It has no direct connections to the outside world (inputs or outputs). All connections from the hidden layer are to other layers within the system.

***Output Layer*:** A layer of neurons that receives processed information and sends output signals out of the system.

***Bias*:** Acts on a neuron like an offset. The function of the bias is to provide a threshold for the activation of neurons. The bias input is connected to each of the hidden and output neurons in a network.

2.6.2 INPUT-OUTPUT MAPPING

The input/output mapping of a network is established according to the weights and the activation functions of their neurons in input, hidden and output layers.

The number of input neurons corresponds to the number of input variables in the neural network, and the number of output neurons is the same as the number of desired output variables. The number of neurons in the hidden layer(s) depends upon the particular NN application. For example, consider the following two-layer feed-forward network with three neurons in the hidden layer and two neurons in the second layer:

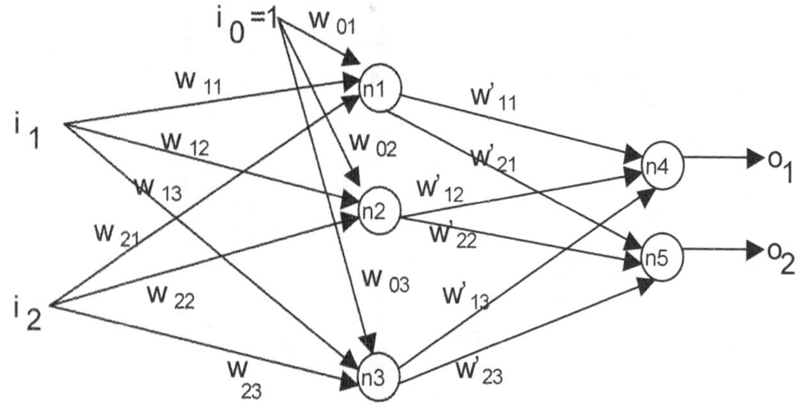

Figure 2.18 An Example of Multi-layer Perceptron.

As is shown, the inputs are connected to each neuron in hidden layer via their corresponding weights. A zero weight indicates no connection. For example, if $W_{23} = 0$, it is implied that no connection exists between the second input (i_2) and the third neuron (n_3). Outputs of the last layer are considered as the outputs of the network.

The structure of each neuron within a layer is similar to the architecture as described in section 2.5. Although the activation function for one neuron could be different from other neurons within a layer, for structural simplicity, similar neurons are commonly chosen within a layer. The input data sets (or sensory information) are presented to the input layer. This layer is connected to the first hidden layer. If there is more than one hidden layer, the last hidden layer should be connected to the output layer of the network. At the first phase, we will have the following linear relationship for each layer:

$$A_1 = W_1 X \qquad .....(2.12)$$

where A_1 is a column vector consisting of m elements, W_1 is an m×n weight matrix and X is a column input vector of dimension n. For the above example, the linear activity level of the hidden layer (neurons n_1 to n_3) can be calculated as follows:

$$\begin{cases} a_{11} = w_{11}i_1 + w_{21}i_2 \\ a_{12} = w_{12}i_1 + w_{22}i_2 \\ a_{13} = w_{13}i_1 + w_{23}i_2 \end{cases} \quad \text{.....(2.13)}$$

The output vector for the hidden layer can be calculated by the following formula:

$$O_1 = F.A_1 \quad \text{.....(2.14)}$$

where A_1 is defined in Equation 2.12, and O_1 is the output column vector of the hidden layer with m element. F is a diagonal matrix comprising the non-linear activation functions of the first hidden layer:

$$F = \begin{bmatrix} f_1(.) & 0 & 0 & \dots & 0 \\ 0 & f_2(.) & & & 0 \\ . & & .. & & .. \\ . & & & .. & 0 \\ 0 & 0 & \dots & 0 & f_m(.) \end{bmatrix} \quad \text{.....(2.15)}$$

For example, if all activation functions for the neurons in the hidden layer of Figure 2.18 are chosen similarly, then the output of the neurons n_1 to n_3 can be calculated as follows:

$$\begin{cases} O_{11} = f(a_{11}) \\ O_{12} = f(a_{12}) \\ O_{13} = f(a_{13}) \end{cases} \quad \text{.....(2.16)}$$

In a similar manner, the output of other hidden layers can be computed. The output of a network with only one hidden layer according to Equation 2.14 is as follows:

$$A_2 = W_2.O_1 \quad \text{.....(2.17)}$$

$$O_2 = G.A_2 \quad \text{.....(2.18)}$$

where A_2 is the vector of activity levels of output layer and O_2 is the q output of the network. G is a diagonal matrix consisting of nonlinear activation functions of the output layer:

$$G = \begin{bmatrix} g_1(.) & 0 & 0 & \dots & 0 \\ 0 & g_2(.) & & & 0 \\ . & & .. & & .. \\ . & & & .. & 0 \\ 0 & 0 & \dots & 0 & g_q(.) \end{bmatrix} \quad \text{....(2.19)}$$

For Figure 2.18, the activity level of output neurons n_4 and n_5 can be calculated as follows:

$$\begin{cases} a_{21} = W'_{11}O_{11} + W'_{12}O_{21} + W'_{13}O_{31} \\ a_{22} = W'_{21}O_{11} + W'_{22}O_{21} + W'_{23}O_{31} \end{cases} \quad \text{.....(2.20)}$$

The two outputs of the network with the similar activation functions can be calculated as follows:

$$\begin{cases} O_1 = g(a_{21}) \\ O_2 = g(a_{22}) \end{cases} \quad \text{.....(2.21)}$$

Therefore, the input-output mapping of a multi-layer perceptron is established according to relationships 2.12–2.22. In sequel, the output of the network can be calculated using such nonlinear mapping and the input data sets.

2.6.3 XOR Realization

As it was shown in section 2.4, a single-layer perceptron cannot classify the input patterns that are not linearly separable such as an Exclusive OR (XOR) gate. This problem may be considered as a special case of a more general non-linear mapping problem. In the XOR problem, we need to consider the four corners of the unit square that correspond to the input pattern. We may solve the problem with a multi-layer perceptron with one hidden layer as shown in Figure 2.19.

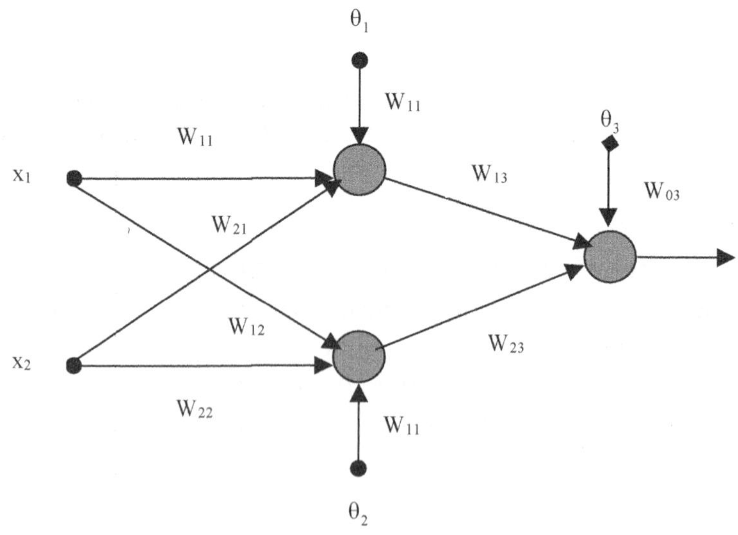

Figure 2.19 Neural Network Architecture to Solve XOR Problem.

In the above configuration, a McCulloh-Pitts model represents each neuron, which uses a hard limit activation function. By appropriate selections of the network weights, the XOR could be implemented using decision surfaces as shown in Figure 2.20.

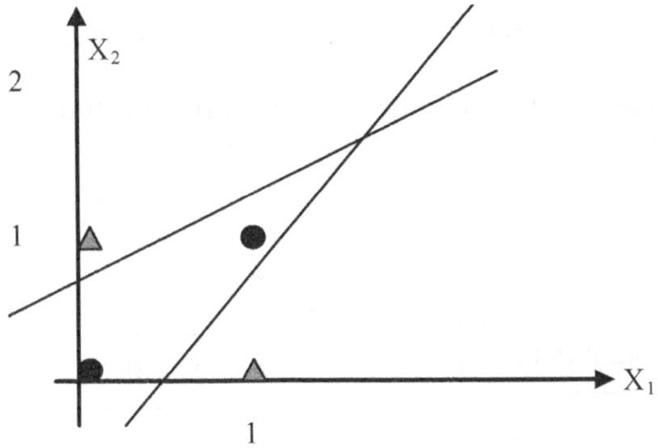

Figure 2.20 Decision Surfaces to Solve XOR Problem.

Example 2.3

Suppose weights and biases are selected as shown in Figure 2.21. The McCulloh-Pitts model represents each neuron (binary hard limit activation function). Show that the network solves XOR problem. In addition, draw the decision boundaries constructed by the network.

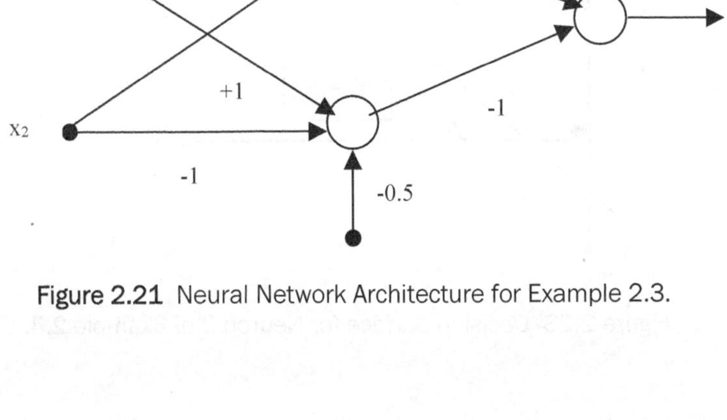

Figure 2.21 Neural Network Architecture for Example 2.3.

In Figure 2.21, suppose the outputs of neurons (before activation function) denote as O_1, O_2, and O_3. The outputs of the summing points at the first layer are as follow:

$$O_1 = x_1 - x_2 + 0.5 \qquad .....(2.22)$$

$$O_2 = x_1 - x_2 - 0.5 \qquad .....(2.23)$$

With the binary hard limited functions, the output y_1 and y_2 are shown in Figures 2.22 and 2.23.

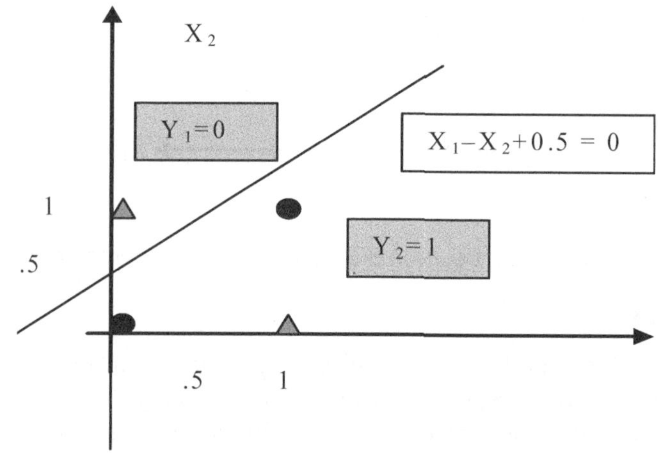

Figure 2.22 Decision Surface for Neuron 1 of Example 2.3.

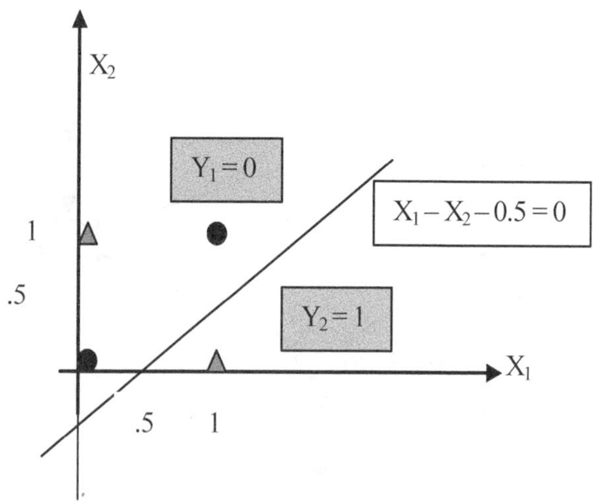

Figure 2.23 Decision Surface for Neuron 2 of Example 2.3.

The outputs of the summing points at the second layer are:

$$O_3 = y_1 - y_2 - 1 \qquad .....(2.24)$$

The decision boundaries of the network are shown in Figure 2.24. Therefore, XOR realization can be accomplished by selection of appropriate weights using Figure 2.19.

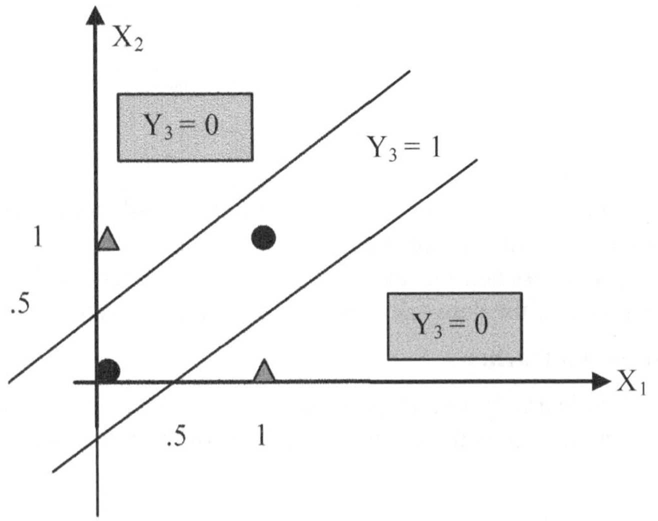

Figure 2.24 Decision Surfaces for Example 2.3.

2.7 LEARNING TYPES

The property that is of primary significance for a neural network is the ability of the network to learn from environment, and to improve its performance through learning.

A neural network learns about its environment through an interactive process of adjustment applied to its synaptic weights and bias levels. Network becomes more knowledgeable about its environment after each iteration of the learning process.

Learning with a teacher:

1. **Supervised learning:** the learning process in which the teacher teaches the network by giving the network the knowledge of environment in the form of sets of the inputs-outputs pre-calculated examples.

As shown in figure

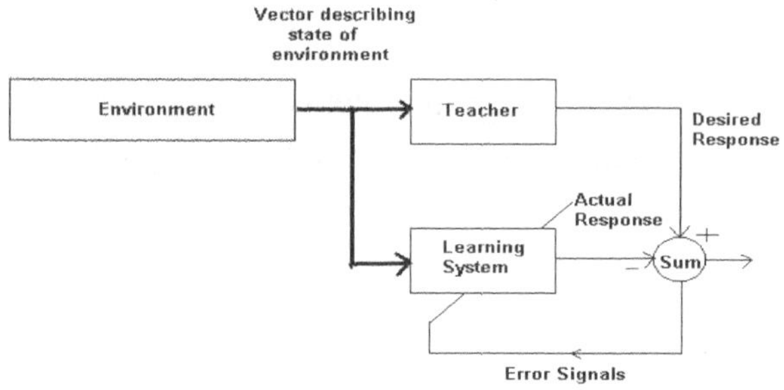

Block Diagram of Supervised Learning

Neural network response to inputs is observed and compared with the predefined output. The difference is calculated refer as "error signal" and that is feed back to input layers neurons along with the inputs to reduce the error to get the perfect response of the network as per the predefined outputs.

Learning without a teacher

Unlike supervised learning, in unsupervised learning, the learning process takes place without teacher that is there are no examples of the functions to be learned by the network.

1. **Reinforcement learning / neurodynamic programming :** In reinforcement learning, the learning of an input output mapping is performed through continued interaction with environment in order to minimize a scalar index of performance.

   As shown in figure.

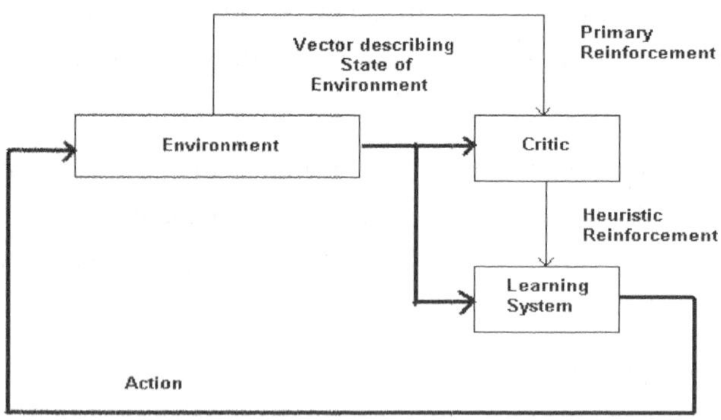

Block Diagram of Reinforcement Learning

In reinforcement learning, because no information on way the right output should be provided, the system must employ some random search strategy so that the space of plausible and rational choices is searched until a correct answer is found. Reinforcement learning is usually involved in exploring a new environment when some knowledge(or subjective feeling) about the right response to environmental inputs is available. The system receives an input from the environment and process an output as response. Subsequently, it receives a reward or a panelty from the environment. The system learns from a sequence of such interactions.

2. ***Unsupervised learning:*** In unsupervised or self-organized learning there is no external teacher or critic to over see the learning process.

   As indicated in figure.

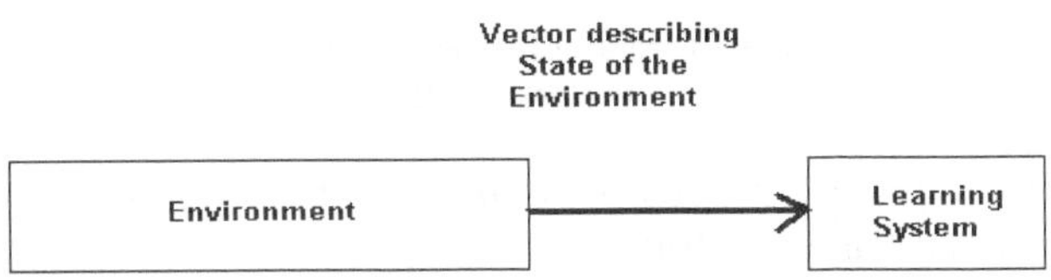

Block Diagram of Unsupervised Learning

Rather provision is made for a task independent measure of the quality of the representation that the network is required to learn and the free parameters of the network are optimized with respect to that measure. Once the network has become tuned to the statistical regularities of the input data, it developes the ability to form internal representation for encoding features of the input and there by to create the new class automatically.

Learning tasks

Pattern recognition: Humans are good at pattern recognition. We can recognize the familiar face of the person even though that person has aged since last encounter, identifying a familiar person by his voice on telephone, or by smelling the fragments comes to know the food etc.

Pattern recognition is formally defined as the process where by a received pattern/signal is assigned to one of a prescribed number of classes. A neural network performs pattern recognition by first undergoing a training session, during which the network is repeatedly present a set of input pattern along with the category to which each particular pattern belongs. Later, a new pattern is presented to the network that has not been seen before, but which belongs to the same pattern caterogy used to train the network.

The network is able to identify the class of that particular pattern because of the information it has extracted from the training data. Pattern recognition performed by neural network is statistical in nature, with the pattern being represented by points in a multidimensional decision space.

The decision space is divided into regions, each one of which is associated with class. The decision boundries are determined by the training process.

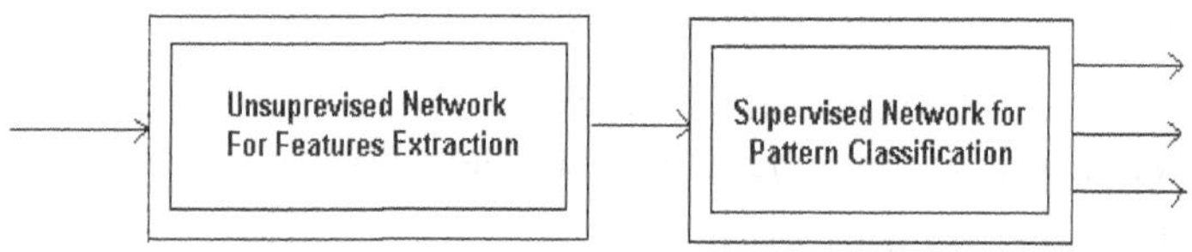

Block Diagram Of Pattern Recognition to classify Pattern

As shown in figure: in generic terms, pattern-recognition machines using neural network may take two forms.

1. To extract features through unsupervised network.
2. Features pass to supervised network for pattern classification to give final output.

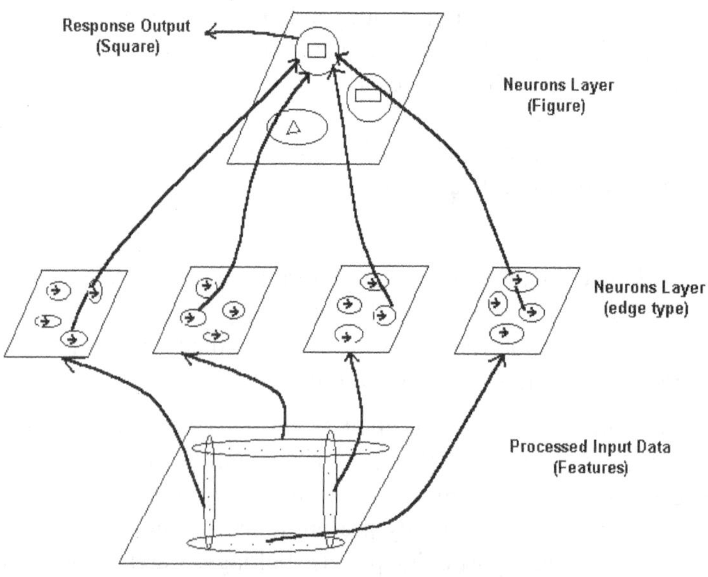

Vision Application How ANN recongnizes

Control

The control of a plant is another learning task that can be done by a neural network; by a 'plant' we mean a process or critical part of a system that is to be maintained in a controlled condition. The relevance of learning to control should not be surprising because, after all, the human brain is a computer, the output of which as a whole system are actions. In the context of control, the brain is living proof that it is possible to build a generalized controller that takes full advantages of parallel distributed hardware, can control many thousands of processes as done by the brain to control the thousands of muscles.

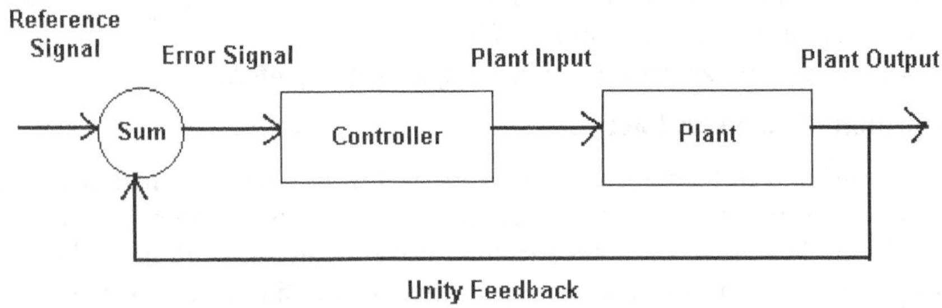

Block Diagram of feedback Control System

Adaptation

The environment of the interest is no stationary, which means that the statistical parameters of the information bearing generated by the environment vary with the time. In situation of the kind, the traditional methods of supervised may learning may prove to be inadequate because the network is not equipped with the necessary means to track the statistical variation of the environment in which it operates. To overcome these shortcomings, it is desirable for a neural network to continually adapt its free parameters to variation in the incoming signals in a real time fashion. Thus an adaptive system responds to every distinct input as a novel one. In other words the learning process encountered in the adaptive system never stops, with learning going on while signal processing is being performed by the system. This form of learning is called continuous learning or learning on the fly.

Generalization

In back propagation learning we typically starts with a training sample and uses the back propagation algorithm to compute the synaptic weights of a multiplayer preceptor by loading (encoding) as many as of the training example as possible into the network. The hope is that the neural network so design will generalize. A network is said generalize well when the input output mapping computed by the network is correct or nearly so for the test data never used in creating or training the network; the term generalization is borrowed from psychology.

A neural network that is design to generalize well will produced a correct input output mapping even when the input is slightly different from the examples used to train the network. When however a neural network learns too many input output examples the network may end up memorizing the training data. It may do so by finding a feature that is present in training data but not true for the underlining function that is to be modeled. Such a phenomena is referred to as an over fitting or over training. When the network is over trained it looses the ability to generalize between similar input output pattern.

The probabilistic neural network

Another multilayer feed forward network is the probabilistic neural network (PNN). In addition to the input layer, the PNN has two hidden layers and an output layer. The major difference from a feed forward network trained by back propagation is that it can be constructed after only a single pass of the training exemplars in its original form and two passes is a modified version. The activation function of a neural in the case of the PNN is statistically derived from estimating of probability density functions (PDFs) based on training patterns.

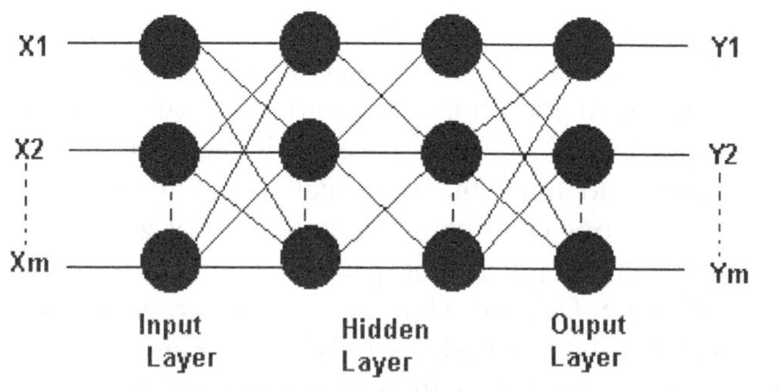

Multilayer Feed Forward Network

CONCLUSION

In this chapter, the fundamentals of neural networks were introduced. The perceptron is the simplest form of neural network used for the classification of linearly separable patterns. Multi-layer perceptron overcome many limitations of single-layer perceptron. They can form arbitrarily complex decision regions in order to separate various nonlinear patterns. The next chapter is devoted to several neural network architectures. (Applications of NN will be presented in Chapters 3-4 and Chapter 8B of the book).

REFERENCES

1. McCulloch, W.W. and Pitts, W., A Logical Calculus of Ideas Imminent in Nervous Activity. *Bull. Math. Biophys.*, 5, 115–133, 1943.
2. Pitts, W. and McCulloch, W.W., How we Know Universals, Bull. *Math.* 127–147, 1947.
3. McClelland, J.L. and Rumelhart, D.E., *Parallel Distributed Processing -Explorations in the Microstructure of Cognition*, Vol. 2, *Psychological and Biological Models*, MIT Press, Cambridge, MA, 1986.
4. Rosenblatt, F., *Principles of Neurodynamics*, Spartan Press, Washington, DC, 1961.
5. Minsky, M. and Papert, S., *Perceptron: An Introduction to Computational Geometry*, MIT Press, Cambridge, MA, 1969.
6. Widrow, B. and Hoff, M.E, Adaptive Switching Circuits, IRE WESCON Convention Record, Part 4, NY, IRE, 96–104, 1960.
7. Fausett, L., *Fundamentals of Neural Networks*, Prentice-Hall, Englewood Cliffs, NJ, 1994.
8. Haykin, S., *Neural Networks: A Comprehensive Foundation*, Prentice Hall, Upper Saddle River, NJ, 1999.
9. Kosko, B., *Neural Network for Signal Processing*, Prentice Hall, Englewood Cliffs, NJ, 1992.
10. Ham, F. and Kostanic, I., *Principles of Neurocomputing for Science and Engineering*, McGraw Hill, New York, NY, 2001.
11. Lippmann, R.P., An Introduction to Computing with Neural Network, *IEEE Acoustic, Speech, and Sig. Proces. Mag.*, 4, 1987.

CHAPTER 3

FEEDFORWARD NEURAL NETWORKS

Introduction

The method of storing and recalling information in brain is not fully understood. However, experimental research has enabled some understanding of how neurons appear to gradually modify their characteristics because of exposure to particular stimuli. The most obvious changes have been observed to occur in the electrical and chemical properties of the synaptic junctions. For example the quantity of chemical transmitter released into the synaptic cleft is increased or reduced, or the response of the postsynaptic neuron to receive transmitter molecules is altered. The overall effect is to modify the significance of nerve impulses reaching that synaptic junction on determining whether the accumulated inputs to post-synaptic neuron will exceed the threshold value and cause it to fire. Thus learning appears to effectively modify the weighting that a particular input has with respect to other inputs to a neuron.

In this chapter, learning in feedforward networks will be considered. Research interest in multilayer feedforward networks dates back to the pioneering work of Rosenblatt (1932) on perceptrons and that of Widrow on Madalines [Widrow 32]. Madalines were constructed with many Adaline elements in the first layer, and with various logic devices such as AND, OR and majority vote-taker elements in the second layer. Madalines of the 1930`s had adaptive first layers and fixed threshold functions in the second (output) layers [Widrow and Lehr 90]. However the tool that was missing in those early days of multilayer feedforward networks was what we now call backpropagation learning.

Usage of the term ***backpropagation*** appears to have evolved in 1985. However, the basic idea of back-propagation was first described by Werbos in his Ph.D. Thesis [Werbos 74], in the context of a more general network. Subsequently, it was rediscovered by Rumelhart, Hinton and Williams (1983b), and popularized through the publication of the seminal book entitled Parallel and Distributed Processing [Rumelhart and McClelland 1983]. A similar generalization of the algorithm was derived independently by Parker, 1985, and interestingly enough, a roughly similar learning algorithm was also studied by LeCun (1985).

3.1 PERCEPTRON CONVERGENCE PROCEDURE

Perceptron was introduced by Frank Rosenblatt in the late 1950's (Rosenblatt, 1958) with a learning algorithm on it. Perceptron may have continuous valued inputs. It works in the same way as the formal artificial neuron defined previously. Its activation is determined by equation:

$$a = \mathbf{w}^{\mathrm{T}}\mathbf{u} + \theta \qquad .....(3.1)$$

Moreover, its output function is:

$$f(a) = \begin{cases} +1 & \text{for } 0 \le a \\ -1 & \text{for } a < 0 \end{cases} \qquad .....(3.2)$$

having value either +1 or –1.

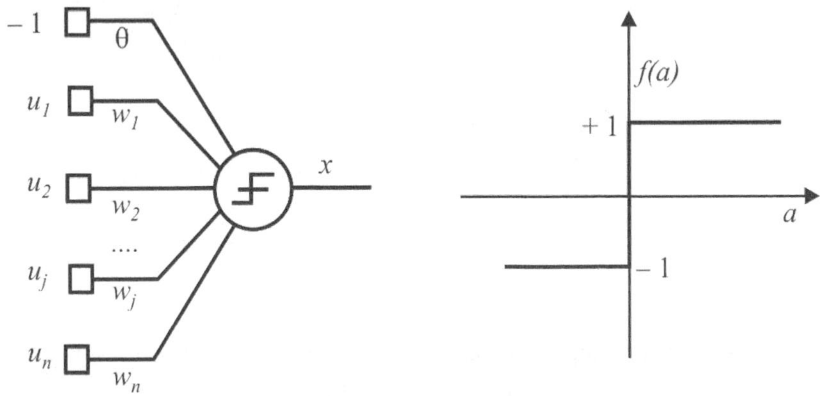

Figure 3.1 Perceptron

Now, consider such a perceptron in N dimensional space (Figure 3.1), the equation

$$\mathbf{w}^{\mathrm{T}}\mathbf{u} + \theta = 0 \qquad .....(3.3)$$

that is

$$w1u1 + w2u2 + ... + wN\ uN + \theta = 0 \qquad .....(3.4)$$

defines a hyperplane. This hyperplane divides the input space into two parts such that at one side, the perceptron has output value +1, and in the other side, it is –1.

A perceptron can be used to decide whether an input vector belongs to one of the two classes, say classes A and B. The decision rule may be set as to respond as class A if the output is +1 and as class B if the output is –1. The perceptron forms two decision regions separated by the hyperplane. The equation of the boundary hyperplane depends on the connection weights and threshold.

Example 3.1

When the input space is two-dimensional then the equation

$$w1u1+w2u2 + \theta = 0 \qquad .....(3.5)$$

defines a line as shown in the Figure 3.2.

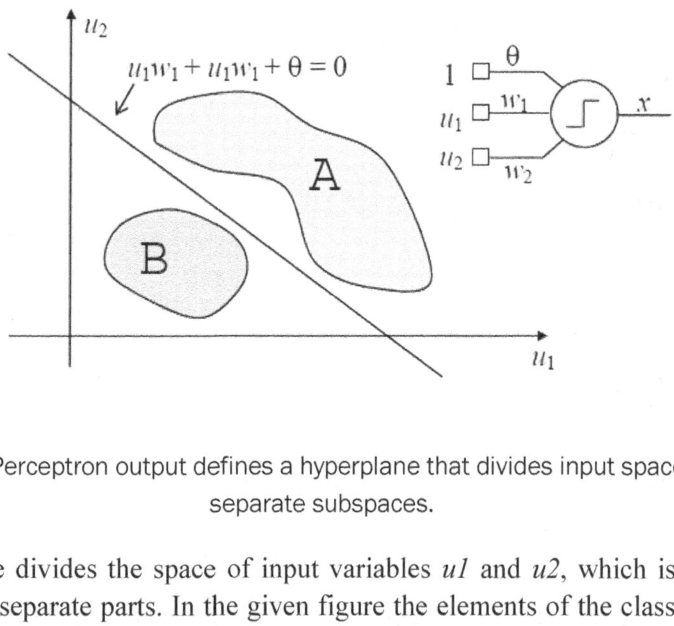

Figure 3.2 Perceptron output defines a hyperplane that divides input space into two separate subspaces.

This line divides the space of input variables *u1* and *u2*, which is a plane, into to two separate parts. In the given figure the elements of the classes A and B lies on the different sides of the line.

Connection weights and the threshold in a perceptron can be fixed or adapted by using a number of different algorithms. The original perceptron convergence procedure developed by Rosenblatt, 1959 for adjusting weights is provided in the following:

The Perceptron Convergence Procedure

Step 1: ***Initialize weights and threshold:*** Set each $wj(0)$, for $j = 0, 1, 2, .., N$, in $\mathbf{w}(0)$ to small random values. Here $\mathbf{w} = \mathbf{w}(t)$ is the weight vector at iteration time t and the component $w0 = \theta$ corresponds to the threshold.

Step 2. ***Present New Input and Desired output:*** Present a new continuous valued input vector $\mathbf{u}k$ to the along with the desired output y^k, such that:

$$y^k = \begin{cases} +1 & for\ u^k \in A \\ -1 & for\ u^k \in B \end{cases}$$

Step 3. Calculate actual output

$$x^k = f\left(\mathbf{w}^{\mathsf{T}}\mathbf{u}^k\right)$$

Step 4. Adapt weights

$$\mathbf{w}\left(t+1 = \mathbf{w}(t) + \eta\left(y^k - x^k(t)\mathbf{u}k\right)\right)$$

where η is a positive constant less than 1.

Step 5. Repeat steps 2-4 until no error occurs

Initially connection weights and bias values are set to small random non-zero values. Then, a new input vector $\mathbf{u}$ with N continuous valued elements is applied to the input and the output value is calculated in Step 2 by using the Eqs. (3.1) and (3.2). Notice that the connection weights are adapted only when an error occurs in step 4 that is when the calculated and the desired values are different. Weights remain unchanged if a correct decision is made by the perceptron. The weight update equation given in this step includes a gain term η that ranges from 0.0 to 1.0 and controls the learning rate. If η is not small enough, then oscillation may occur during weight adaptation. On the other hand, if η is too small then adaptation rate is very slow.

Example 3.2

Figure 3.3 demonstrates how the line defined by the perceptrons parameters is shifted in time as the weights are updated. Although it is not able to separate the classes A and B with the initial weights assigned at time $t = 0$, it manages to separate them at the end.

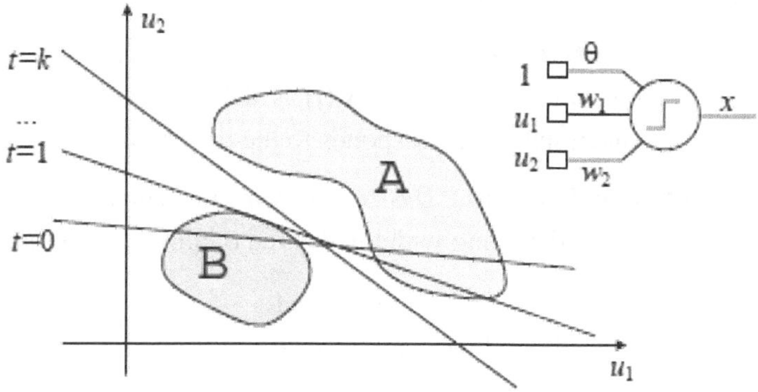

Figure 3.3 Perceptron convergence.

[Rosenblatt, 1959] it is proved that if the inputs presented from the two classes are separable, that is if they fall on the opposite sides of some hyperplane, then the perceptron convergence procedure always converge in time. Furthermore, it positions the final decision hyperplane such that it separates the samples of class A from those of class B.

One problem with the perceptron convergence procedure is that the decision boundary may oscillate continuously when the distributions overlap or the classes are not linearly separable (Figure 3.4).

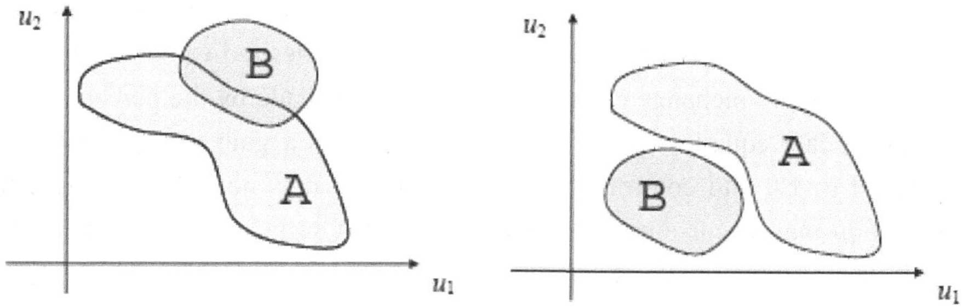

Figure 3.4 (a) Overlapping distributions (b) non linearly separable distribution.

The types of decision regions that can be formed by single and multilayer perceptrons with one and two layers of hidden layers are given in the Figure 3.5 [Lipmann 87].

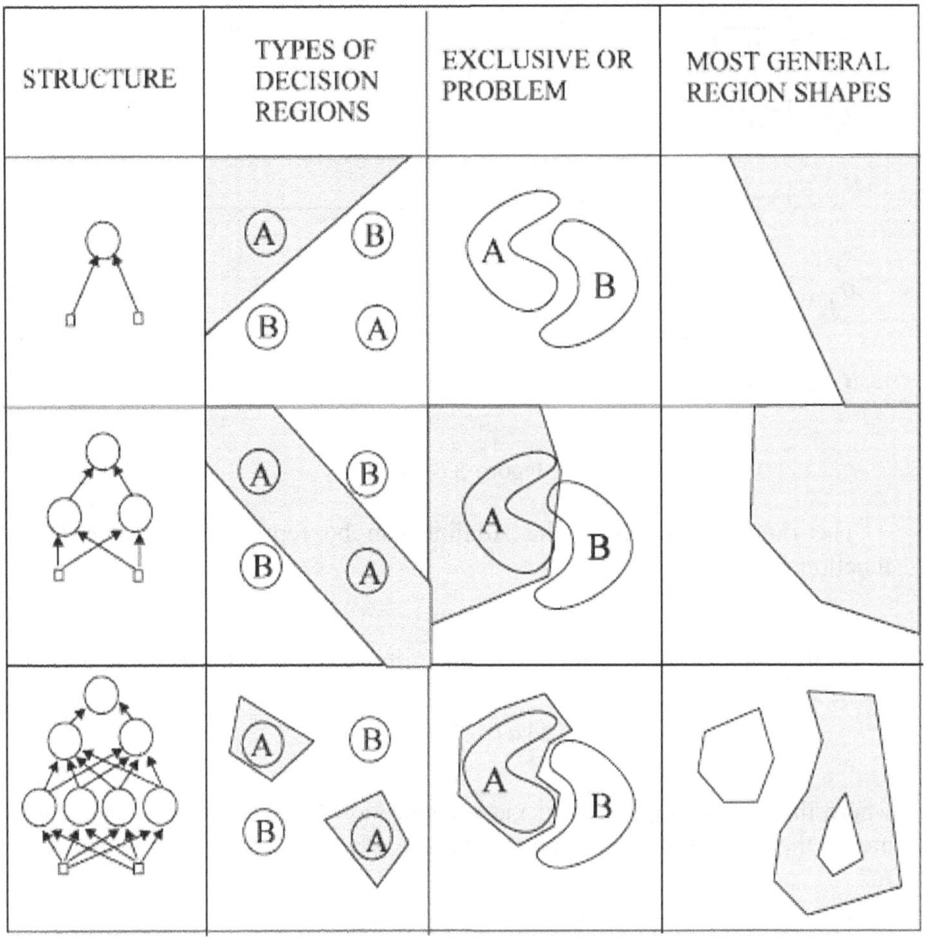

Figure 3.5 Types of regions that can be formed by single and multi-layer perceptrons (Adapted from Lippmann 87).

3.2 LMS LEARNING RULE

A modification to the perceptron convergence procedure forms the Least Mean Square (LMS) solution for the case that the classes are not separable. This solution minimizes the mean square error between the desired output and the actual output of the processing element. The LMS algorithm was first proposed for Adaline (Adaptive Linear Element) in [Widrow and Hoff 30]. The structure of Adaline is shown in the Figure 3.6. The part of the Adaline that executes the summation is called Adaptive Linear Combiner.

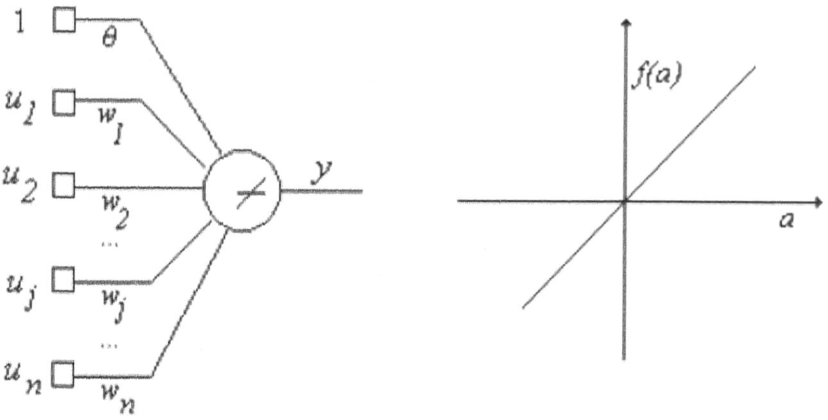

Figure 3.6 Adaline.

The output function of the Adaline can be represented by the identity function as:

$$f(a) = a, \qquad .....(3.6)$$

So the output can be written in terms of input and weights as:

$$x = f(a) = \sum_{j=0}^{N} w_j u_j \qquad .....(3.7)$$

where the bias is implemented via a connection to a constant input $u0$, which means that the input vector and the weight vector are of space $R^{(N+1)}$ instead of R^N.

The output equation of Adaline can be written as:

$$x = \mathbf{w}^{\mathrm{T}}\mathbf{u} \qquad .....(3.8)$$

where **w** and **u** are weight and input vectors respectively having dimension $N + 1$.

Suppose that we have a set of input vectors $\mathbf{U}^k$, $k = 1..\ K$, each having its own desired output value Y^k.

The performance of the Adaline for a given input value U^k, can be defined by considering the difference between the desired output Y^k and the actual output X^K, which is called error and denoted as ε. Therefore, the error for the input U^k, is as follows:

$$\varepsilon^k = y^k - x^k = y^k - \mathbf{w}^{\mathrm{T}}\mathbf{u}^k \qquad .....(3.9)$$

The aim of the LMS learning is to adjust the weights through a training set $\{(\mathbf{U}^k, Y^k)\}$, $k = 1..K$, such that the mean of the square of the errors is minimum. The mean square error is defined as:

$$<(\varepsilon^k)^2> \ = \ \lim_{k\to\infty} 1/K \sum_{k=1}^{K} (\varepsilon^k)^2 \qquad .....(3.10)$$

where the notation <.> denotes the mean value.

The mean square error can be rewritten as:

$$<(\varepsilon^k)^2> = <(y^k - \mathbf{w}^{\mathsf{T}}\mathbf{u}^k)^2>$$
$$= <(y^k)^2> + \mathbf{w}^{\mathsf{T}} <\mathbf{u}^k \times \mathbf{u}^k> \mathbf{w} - 2 <y^k \mathbf{u}^{k\mathsf{T}}> \mathbf{w} \qquad .....(3.11)$$

where T denotes transpose and x is the outer vector product.

Defining input correlation matrix **R** [Widrow 85, Freeman 91]

$$\mathrm{R} = <\mathbf{u}^k \times \mathbf{u}^k> = <\mathbf{u}^k\, \mathbf{u}^{k\mathsf{T}}> \qquad .....(3.12)$$

and a vector **P** as

$$\mathrm{P} = <y^k\, \mathbf{u}^k> \qquad .....(3.13)$$

results in:

$$\mathsf{e}(\mathbf{w}) = <(\varepsilon^k)^2> = <(y^k)^2> + \mathbf{w}^{\mathsf{T}}\mathbf{R}\,\mathbf{w} - 2\mathbf{P}^{\mathsf{T}}\mathbf{w} \qquad .....(3.14)$$

The optimum value **w*** for the weight vector corresponding to the minimum of the mean squared error can be obtained by evaluating the gradient of e(**w**). The point which makes the gradient zero gives us the value of **w***, that is:

$$\nabla\mathsf{e}(\mathbf{w})\Big|_{\mathbf{w}=\mathbf{w}^*} = \frac{\partial\mathsf{e}(\mathbf{w})}{\partial\mathbf{w}}\Bigg|_{\mathbf{w}=\mathbf{w}^*} = 2\mathbf{R}\mathbf{w}^* - 2\mathbf{P} = 0 \qquad .....(3.15)$$

Here, the Gradient is,

$$\Delta\mathsf{e}(\mathbf{w}) = \left[\partial\mathsf{e}/\partial w_1 \quad \partial\mathsf{e}/\partial w_2 \quad \cdots \quad \partial\mathsf{e}/\partial w_n\right]^{\mathsf{T}} \qquad .....(3.16)$$

and it is a vector extending in the direction of the greatest rate of change. The gradient of a function evaluated at some point is zero if the function has a maximum or minimum at that point. The error function is of the second degree so it is a paraboloid and it has a single minimum at point **w***.

When we set the gradient of the mean square error to zero, this implies that

$$\mathbf{R}\mathbf{w}^* = \mathbf{P} \qquad .....(3.17)$$

And then,

$$\mathbf{w}^* = \mathbf{R}^{-1}\mathbf{P} \qquad .....(3.18)$$

3.3 STEEPEST DESCENT ALGORITHM

The analytical calculation of the optimum weight vector for a problem is rather difficult in general. Not only does the matrix manipulation get cumbersome for the large dimensions, but also each component of **R** and **P** itself is an expectation value. Thus, explicit calculations of **R** and **P** require knowledge of the statistics of the input signal [Freeman 91]. A better approach would be to let the Adaline Linear Combiner to find the optimum weights by itself through a search over the error surface. Instead of having a purely random search, some intelligence is added to the procedure such that the weight vector is changed by considering the gradient of e(**w**) iteratively [Widrow 30], according to formula known as *delta rule*:

$$\mathbf{w}(t+1) = \mathbf{w}(t) + \Delta\mathbf{w}(t) \qquad \text{.....(3.19)}$$

where

$$\Delta\mathbf{w}(t) = -\eta \nabla e(\mathbf{w}(t)) \qquad \text{.....(3.20)}$$

In the above formula η is a small positive constant, determining the learning rate. For the real valued scalar function e (**w**) on a vector space **w** *RN*, the gradient e(**w**) gives the direction of the steepest upward slope, so the negative of the gradient is the direction of the steepest descent . This fact is demonstrated in Figure 3.7 for a parabolic error surface on two dimensions.

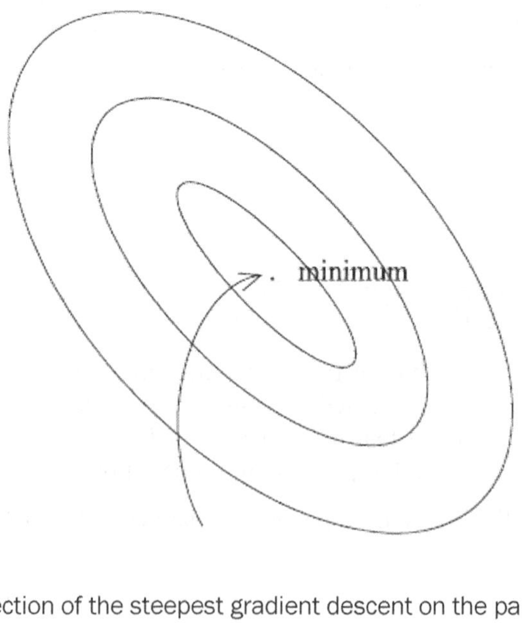

Figure 3.7 Direction of the steepest gradient descent on the paraboloid error surface on two-dimensional weight space. Only the equipotential curves of the error surface is shown instead of the 3D-error surface.

In Section 3.2 we have considered the linear output function in the derivation of the optimum weight **w*** for the minimum error. However in the general case, we should consider any nonlinearity $f(.)$ at the output of the neuron. It should be noted that in such a case the error surface is no more a paraboloid, so it may have several local minima. For an input $\mathbf{U}^k$ applied at time t, $(\varepsilon^k(t))^2$ can be used as an approximation to $<(\varepsilon^k)>$,

where

$$\varepsilon^k(t) = y^k - f(a^k) = y^k - f(\mathbf{w}(t)^{\mathsf{T}}\mathbf{u}^k) \qquad .....(3.21)$$

Therefore, we obtain:

$$\nabla < (\varepsilon^k)^2 > \;\cong\; \nabla(\varepsilon^k(t))^2 = (y^k - f(a^k))^2 \qquad .....(3.22)$$

With a differentiable function $f(.)$, having derivative $f'(.)$ it becomes

$$\nabla(y^k - f(a^k))^2 = -2\varepsilon^k(t) f'(a^k)\nabla a^k \qquad .....(3.23)$$

since

$$\nabla a^k = \nabla\mathbf{w}(t)^{\mathsf{T}}\mathbf{u}^k = \mathbf{u}^k \qquad .....(3.24)$$

Then the weight update formula becomes:

$$\mathbf{w}(t+1) = \mathbf{w}(t) + 2\eta\varepsilon^k(t) f'(a^k)\mathbf{u}^k \qquad .....(3.25)$$

Notice that for Adaline's linear output function:

$$f'(a) = 1 \qquad .....(3.26)$$

For sigmoid function it is:

$$f'(a) = \frac{\partial}{\partial a}\left(\frac{1}{1+e^{-a/T}}\right) = \frac{1}{T} f(a)(1 - f(a)) \qquad .....(3.27)$$

The steepest descent algorithm based on least mean square error is summarized in the following:

Steepest descent algorithm

Step 1: Apply an input vector $\mathbf{u}^k$ with an desired output value y^k to the neuron's inputs.

Step 2: By considering $\mathbf{u}^k$ and using the current value of the weight vector determine the value of the activation a^k:

$$a^k = \mathbf{w}(t)^{\mathsf{T}}\mathbf{u}^k$$

Step 3: Determine the value of the derivative of the output function using the current value of activation a^k, that is:

$$f(a^k) = \left.\frac{\partial f(a)}{\partial a}\right|_{a=a^k}$$

Step 4: Determine the value of error $\varepsilon^k(t)$ as: $\varepsilon^k(t) = y^k - f(a^k)$

Step 5: Update the weight vector using the following update formula

$$\mathbf{w}(t+1) = \mathbf{w}(t) + 2\eta\, f'\,(a^k)\varepsilon^k(t)\,\mathbf{u}^k$$

Step 6: Repeat steps 1-5 until $< \varepsilon^k(t)^2 >$ reduces to an acceptable level.

The parameter η in the algorithm determines the stability and the speed of convergence of the weight vector towards the minimum error value. The value of η should be tuned well. If it is chosen too small this effects considerably the convergence time. On the other hand, if changes are too large, the weight vector may wander around the minimum as shown in the Figure 3.8, without being able to reach it.

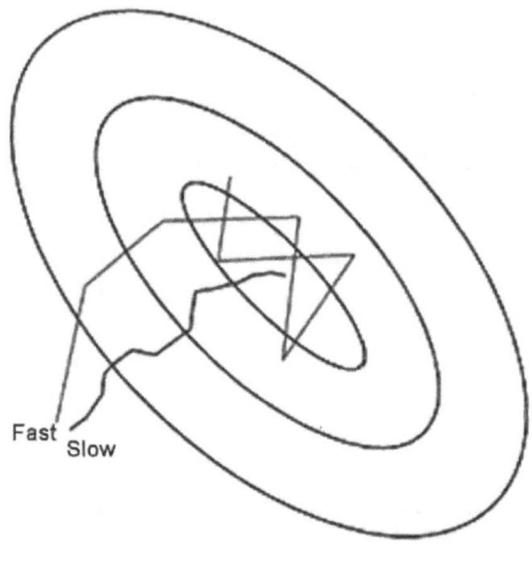

Figure 3.8 Inappropriate value of learning rate η may cause oscillations in the weight values without convergence.

Notice that, the iterative weight update by the delta rule is derived by assuming constant $\mathbf{U}^k$. Therefore, it tends to minimize the error with respect to applied $\mathbf{U}^k$. In fact, we require the average error, that is:

$$\mathsf{e} = <(\varepsilon^k)^2> = \frac{1}{K}\sum_{k=1}^{K}(\varepsilon^k)^2 \quad .....(3.28)$$

to be minimum, and this implies that

$$\frac{\partial \mathsf{e}}{\partial w_j} = \frac{1}{K}\sum_{k=1}^{K}\frac{\partial(\varepsilon^k)^2}{w_j} = \frac{1}{K}\sum_{k=1}^{K}\frac{2\varepsilon^k \partial\varepsilon^k}{w_j} \quad .....(3.29)$$

Therefore, the net change in *wj* after one complete cycle of pattern presentation is expected to be:

$$w_j(t+K) = w_j(t) - \eta \frac{1}{K} \sum_{k=1}^{K} \frac{2\varepsilon^k \partial \varepsilon^k}{\partial wj} \quad .....(3.30)$$

However, this would be true that if the weights are not updated along a cycle, but only at the end. By changing the weights as each pattern is presented, we depart to some extend from the gradient descent in e. Nevertheless, provided the learning rate is sufficiently small, this departure will be negligible and the delta rule will implement a very close approximation to gradient descent in mean squared error [Freeman 91].

3.4 THE BACKPROPAGATION ALGORITHM

3.4.1 LEARNING SINGLE LAYER NETWORK

Consider a single layer multiple output network as shown in the Figure 3.9. Here, we still have *N* inputs denoted *uj*, $j = 1..N$, but *M* processing elements whose activations and outputs are denoted as *ai* and *xi* , $I = 1..M$ respectively. Here *wji* is used to denote the strength of the connection from the *j*th input to the *i*th processing element. In vector notation *Wji* is the jth component of weight vector **w***i*, while $\mathbf{U}^k$ is the *j*th component of the input vector **u.** Let $\mathbf{U}^k$ and Y^k to represent the *k*th input sample and the corresponding desired output vector respectively.

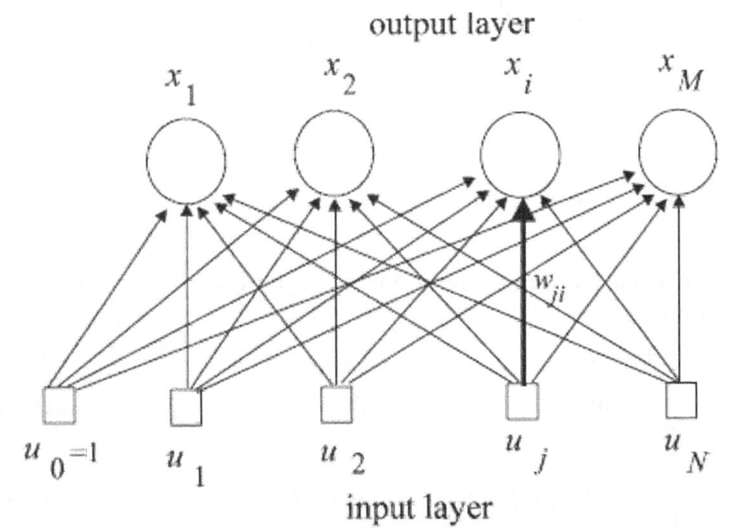

Figure 3.9 Multiple output network.

Let the error observed at the output i be

$$\varepsilon_i^k = y_i^k - x_i^k \qquad .....(3.31)$$

when $\mathbf{U^k}$ is applied at the input. If the error is to be written in terms of the input vector $\mathbf{U^k}$ and the weights $\mathbf{w}i$, we obtain

$$\varepsilon_i^k = y_i^k - f(\mathbf{w}_i{}^{\mathsf{T}}\mathbf{u}^k) \qquad .....(3.32)$$

If we take partial derivative with respect to wji by applying the chain rule

$$\frac{\partial \varepsilon_i^k}{\partial w_{ji}} = \frac{\partial \varepsilon_i^k}{\partial x_i^k}\frac{\partial x_i^k}{\partial w_{ji}} \qquad .....(3.33)$$

where
$$\frac{\partial \varepsilon_i^k}{\partial x_i^k} = -1 \qquad .....(3.34)$$

and
$$\frac{\partial \varepsilon_i^k}{\partial w_{ji}} = f'(a_i^k)u_j^k \qquad .....(3.35)$$

We obtain
$$\frac{\partial \varepsilon^k}{\partial w_{ji}} = f'(a_i^k)u_j^k \qquad .....(3.36)$$

If we define the total output error for input $\mathbf{U^k}$ as the sum of the square of the errors at each neuron output, that is:

$$\frac{\partial \mathrm{e}^k}{\partial w_{ji}} = \frac{\partial \mathrm{e}^k}{\partial \varepsilon_i^k}\frac{\partial \varepsilon_i^k}{\partial w_{ji}} \qquad .....(3.37)$$

Which is
$$\frac{\partial \mathrm{e}^k}{\partial w_{ji}} = -\varepsilon_i^k f'(a^k)u_j \qquad .....(3.38)$$

By defining $\delta_i^k = \varepsilon_i^k f'\left(a^k\right)$ .....(3.39)

it can be reformulated as

$$\frac{\partial \mathrm{e}^k}{\partial w_{ji}} = -\delta_i^k u_j^k \qquad .....(3.40)$$

For the error to be minimum, the gradient of the total error with respect to the weights should be

$$\nabla \mathrm{e}^k = 0 \qquad .....(3.41)$$

where $\mathbf{0}$ is the vector having $N.M$ entries each having value zero. In other words, it should be satisfied:

$$\frac{\partial \mathrm{e}^k}{\partial w_{ji}} = 0 \quad for\, j = 1..N, \qquad i = 1..M \qquad .....(3.42)$$

In order to reach the minimum of the total error, without solving the above equation, we apply the delta rule in the same way explained for the steepest descent algorithm:

$$\mathbf{w}(t+1) = \mathbf{w}(t) - \eta\, e^k \qquad .....(3.43)$$

in which

$$w_{ji}(t+1) = w_{ji}(t) - \eta \frac{\partial e^k}{\partial w_{ji}} \qquad for\; j = 1..N, \quad i = 1..M \qquad .....(3.44)$$

that is

$$w_{ji}(t+1) = w_{ji}(t) + \eta \delta_i^k u_j^k \qquad for\; j = 1..N, \quad i = 1..M \qquad .....(3.45)$$

3.4.2 MULTILAYER NETWORK

Now assume that another layer of neurons is connected to the input side of the output layer. Therefore we have the input, hidden and the output layers as shown in Figure 3.10. In order to discriminate between the elements of the hidden and output layers we will use the subscripts L and o respectively. Furthermore, we will use h as the index on the hidden layer elements, while still using index j and i for the input and output layers.

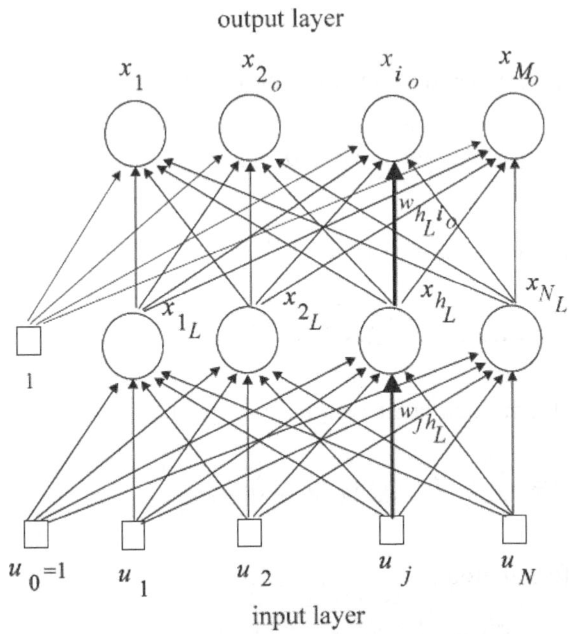

Figure 3.10 Multilayer network.

In such a network, the output value of ith neuron of output layer can be written as:

$$x_{i_o}^k - f_o(w_{i_o}{}^{\mathsf{T}} x_L^k) \qquad .....(3.46)$$

where x_L^k being the vector of output values at hidden layer that is connected as input to the output layer. The value of the hth element in x_L^k is determined by the equation:

$$x_{h_L}^k - f_L(\mathbf{w}_{h_L}{}^{\mathsf{T}} \mathbf{u}^k) \qquad ......(3.47)$$

since

$$\mathbf{w}_{h_L}{}^{\mathsf{T}} \mathbf{u}^k = \sum_{j=1}^{N} w_{jh_L} u_j^k \qquad ......(3.48)$$

the partial derivative of the output of a neuron *io* of output layer with respect to hidden layer weight *wjhL* can be determined by applying the chain rule

$$\frac{\partial x_{i_o}^k}{\partial w_{jh_L}} = \frac{\partial x_{i_o}^k}{\partial x_{h_L}^k} \frac{\partial x_{h_L}^k}{\partial w_{ij_L}} \qquad .....(3.49)$$

By using Eq. (3.46) and (3.48) this can be written as

$$\frac{\partial x_{i_o}^k}{\partial w_{jh_L}} = (f'_o(a_{i_o}^k) w_{h_{Li_o}})(f'_L(a_{h_L}^k) u_j^k) \qquad .....(3.50)$$

Then the partial derivative of the total error

$$\mathrm{e}^k = \tfrac{1}{2} \sum_{i_o=1}^{M} (\varepsilon_{i_o}^k)^2 = \tfrac{1}{2} \sum_{i_o=1}^{M} \left(y_{i_o}^k - x_{i_o}^k\right)^2 \qquad ......(3.51)$$

with respect to the hidden layer weight *wjhL* it can be written as

$$\frac{\partial \mathrm{e}^k}{\partial w_{jh_L}} = -\sum_{i_o=1}^{M} \varepsilon_{i_o}^k f_o'(a_{i_o}^k) w_{h_L i_o} f_L'(a_{hL}^k) u_j^k \qquad ......(3.52)$$

It can be reformulated as

$$\frac{\partial \mathrm{e}^k}{\partial w_{jh_L}} = -\sum_{i_o=1}^{M} \delta_{i_o}^k w_{h_L i_o} f_L'(a_{hL}^k) u_j^k \qquad .....(3.53)$$

When defined

$$\delta_{h_L}^k = f_L'\left(a_{h_L}^k\right)\sum_{i_o=1}^{M}\delta_{i_o}^k w_{h_L i_o} \qquad .....(3.54)$$

It becomes

$$\frac{\partial e^k}{\partial w_{jh_L}} = -\delta_{h_L}^k u_j \qquad ......(3.55)$$

Therefore, the weight update rule for the hidden layer

$$w_{jh_L}(t+1) = w_{jh_L}(t) - \eta\frac{\partial e^k}{\partial w_{jh_L}} \qquad ......(3.56)$$

can be reformulated in analogy with the weight update rule of the output layer, as

$$w_{jh_L}(t+1) = w_{jh_L}(t) + \eta\delta_{hL}^k u_j \qquad .....(3.57)$$

This weight update rule may be generalized for the networks having several hidden layers as:

$$w_{j(L-1)h_L}(t+1) = w_{j(L-1)h_L}(t) + \eta\delta_{hL}^k x_{j(L-1)} \qquad .....(3.58)$$

where L and (L-1) are used to denote any hidden layer and its previous layer respectively.

Furthermore,

$$\delta_{j(L-1)}^k = f_{L-1}'(a_{j(L-1)}^k)\sum_{h_L=1}^{N_L}\delta_{h_L}^k w_{j(t-1)h_L} \qquad .....(3.59)$$

where NL is the number of neurons at layer L.

The backpropagation algorithm for multi-layered network is summarized in the following.

Backpropagation Algorithm for Multilayered Feedforward Neural Network

Step 0: Intialize weights: To small random values:

Step 1: Apply a sample: Apply to the input a sample vector $\mathbf{u}^k$ having desired output vector y^k;

Step 2: Forward phase: Starting from the first hidden layer and propagating towards the output layer:

2.1 Calculate the activation values for the units at layer L as:

2.1.1 If L –1 is the input layer

$$a_{h_L}^k = \sum_{j=0}^{N} w_{jh_L} u_j^k$$

2.1.2 If L – 1 is a hidden layer

$$a_{h_L}^k = \sum_{j=0}^{N_{i=1}} w_{j(L-1)} h_L x_{j(L-1)}^k$$

2.2 Calculate the output values for the units at Layer L as:

$$x_{h_L}^k = f_L(a_{h_L}^k)$$

in which use i_o instead of h_L if it is an output layer.

Step 4: Output errors: Calculate the error terms at the output layer as:

$$\delta_{i_o}^k = (y_{i_o}^k - x_{i_o}^k) f_a^{'}(a_{i_o}^k)$$

Step 5: Backward phase propagate error backward to the input layer through each layer L using the error term

$$\delta_{h_L}^k = f_L^{'}(a_{h_L}^k) \sum_{i_{L+1}-1}^{N_{i=1}} \delta_{i(L+1)}^k w_{h_L(K-1)}^k$$

in which, use i_o instead of $i_{(L+1)}$ if $L + 1$ is an output layer.

Step 6: Weight update: Update weights according to the formula

$$w_{j(L-1)h_L}(t+1) = w_{j(L-1)h_L}(t) + \nu = \eta \delta_{h_L}^k x_{j(L-1)}^k$$

Step 7: Repeat steps 1-6 until the stop criterion is satisfied, which may be chosen as the mean of the total error

$$< e^k > \; = \; < \tfrac{1}{2} \sum_{t-1}^{M} (y_{i_o}^k - x_{i_o}^k)^2 >$$

is sufficiently small.

3.5 THE BACK-PROPAGATION ALGORITHM – A MATHEMATICAL APPROACH

Units are connected to one another. Connections correspond to the edges of the underlying directed graph. There is a real number associated with each connection, which is called the weight of the connection. We denote by Wij the weight of the connection from unit ui to unit uj. It is then convenient to represent the pattern of connectivity in the network by a weight matrix W whose elements are the weights Wij. Two types of connection are usually distinguished: excitatory and inhibitory. A positive weight represents an excitatory connection whereas a negative weight represents an inhibitory connection. The pattern of connectivity characterises the architecture of the network.

First, it computes the total weighted input xj, using the formula:

$$X_j = \sum_i y_i W_{ij}$$

where y_i is the activity level of the jth unit in the previous layer and Wij is the weight of the connection between the ith and the jth unit.

Next, the unit calculates the activity y_j using some function of the total weighted input. Typically we use the sigmoid function:

$$y_j = \frac{1}{1+e^{-x_1}}$$

Once the activities of all output units have been determined, the network computes the error E, which is defined by the expression:

$$E = \frac{1}{2}\sum_i (y_i - d_i)^2$$

where y_j is the activity level of the jth unit in the top layer and d_j is the desired output of the jth unit.

The back-propagation algorithm consists of four steps:

1. Compute how fast the error changes as the activity of an output unit is changed. This error derivative (EA) is the difference between the actual and the desired activity.

$$EA_j = \frac{\partial E}{\partial y_i} = y_j - d_j$$

2. Compute how fast the error changes as the total input received by an output unit is changed. This quantity (EI) is the answer from step 1 multiplied by the rate at which the output of a unit changes as its total input is changed.

$$EI_j = \frac{\partial E}{\partial x_j} = \frac{\partial E}{\partial y_j} \times \frac{dy_j}{dx_j} = EA_j y_j (1 - y_j)$$

3. Compute how fast the error changes as a weight on the connection into an output unit is changed. This quantity (EW) is the answer from step 2

multiplied by the activity level of the unit from which the connection emanates.

$$EW_{ij} = \frac{\partial E}{\partial W_{ij}} = \frac{\partial E}{\partial k_j} \times \frac{dk_j}{dW_{ij}} = EI_j y_i (1 - y_j)$$

4. Compute how fast the error changes as the activity of a unit in the previous layer is changed. This crucial step allows back propagation to be applied to multilayer networks. When the activity of a unit in the previous layer changes, it affects the activites of all the output units to which it is connected. So to compute the overall effect on the error, we add together all these seperate effects on output units. But each effect is simple to calculate. It is the answer in step 2 multiplied by the weight on the connection to that output unit.

$$EA_i = \frac{\partial E}{\partial y_i} = \sum_j \frac{\partial E}{\partial x_j} \times \frac{\partial x_j}{\partial y_{ij}} = \sum_j EI_j W_{ij}$$

By using steps 2 and 4, we can convert the EAs of one layer of units into EAs for the previous layer. This procedure can be repeated to get the EAs for as many previous layers as desired. Once we know the EA of a unit, we can use steps 2 and 3 to compute the EWs on its incoming connections.

CONCLUSIONS

In this chapter, we briefly reviewed the feedforward type of neural networks, which are exemplified by multilayer perceptrons (MLPs), radial-basis function (RBF) networks, principal component analysis (PCA) networks, and Perceptron Convergence Theorem. The training of MLPs and RBF networks proceeds in a supervised manner, whereas the training of PCA networks and SOMs proceeds in an unsupervised manner. Feedforward networks by themselves are nonlinear static networks. They can be made to operate as nonlinear dynamical systems by incorporating short-term memory into their input layer. Two important examples of short-term memory are the standard tapped delay-line and the gamma memory that provides control over attainable memory depth. The attractive feature of nonlinear dynamical systems built in this way is that they are inherently stable.

REFERENCES

1. Anderson, J. A., 1995, *Introduction to Neural Networks* (Cambridge, MA: MIT Press).

2. Barlow, H. B., 1989, "Unsupervised learning," *Neural Computation*, vol. 1, pp. 295–311.

3. Becker, S., and G. E. Hinton, 1982, "A self-organizing neural network that discovers surfaces in random-dot stereograms," *Nature (London)*, vol. 355, pp. 161–163.

4. Broomhead, D. S., and D. Lowe, 1988, "Multivariable functional interpolation and adaptive networks," *Complex Systems*, vol. 2, pp. 321–355.

5. Comon,P., 1994,"Independent component analysis:A new concept?" *Signal Processing*, vol. 36, pp. 287–314.

6. deVries, B., and J. C. Principe, 1992, "The gamma model—A new neural model for temporal processing," *Neural Networks*, vol. 4, pp.565–576.

7. Haykin, S., 1999, *Neural Networks: A Comprehensive Foundation*, 2nd ed. (Englewood Cliffs, NJ: Prentice-Hall).

8. Moody and Darken, 1989, "Fast learning in networks of locally-tuned processing unites," *Neural Computation*, vol. 1, pp. 281–294.

9. Oja, E., 1982, "A simplified neuron model as a principal component analyzer," *J. Math. Biol.*, vol. 15, pp. 267–273.

10. Park, J., and Sandberg, I.W., 1993, "Approximation and radial-basis function networks," *Neural computation*, vol. 5, pp. 305–316.

11. Poggio,T., and F. Girosi, 1990, "Networks for approximation and learning," *Proc. IEEE*, vol. 78, pp. 1481–1497.

12. Rumelhart, D. E., G. E. Hinton, and R. J.Williams, 1986, "Learning internal representations by error propagation," in D. E. Rumelhart and J. L.McCleland, eds. (Cambridge, MA: MIT Press), vol. 1, Chapter 8.

13. Sandberg, I.W., 1991a, "Structure theorems for nonlinear systems," *Multidimensional Sys. Sig. Process.* vol. 2, pp. 267–286. (Errata in 1992, vol. 3,p. 101.)

14. Sandberg, I. W., 1991b, "Approximation theorems for discrete-time systems," *IEEE Trans. Circuits Sys.* vol. 38, no. 5, pp. 564–566, May 1991.

CHAPTER 4

NEURAL NETWORKS ARCHITECTURES

4.1 INTRODUCTION

Interest in the study of neural networks has grown remarkably in the last two decades. This is due to the conceptual viewpoint regarding the human brain as a model of a parallel computation device, a very different one from a traditional serial computer. Neural networks are commonly classified by their network topology, node characteristics, learning, or training algorithms. On the other hand, the potential benefits of neural networks extend beyond the high computation rates provided by massive parallelism of the networks. They typically provide a greater degree of robustness or fault tolerance than Von Neumann sequential computers. Additionally, adaptation and continuous learning are integrated components of NN. These properties are very beneficial in areas where the training data sets are limited or the processes are highly nonlinear. Furthermore, designing artificial neural networks to solve problems and studying real biological networks, this chapter may also change the way we think about the problems and may lead us to new insights and algorithm improvements.

The main goal of this chapter is to provide the readers with the conceptual overviews of several neural network architectures. The chapter will not delve too deeply into the theoretical considerations of any one network, but will concentrate on the mechanism of their operation. Examples are provided for each network to clarify the described algorithms and demonstrate the reliability of the network. In the following four chapters various applications pertaining to these networks will be discussed.

This chapter is organized as follows. In section 4.2, various classifications of neural networks according to their operations and/or structures are presented. Feedforward and feedback networks are discussed. Furthermore, two different methods of training, namely supervised and unsupervised learning, are described. Section 4.4 is devoted to error back propagation (BP) algorithm. Various properties of this network are also discussed in this section. Radial basis function network (RBFN) is a feedforward network with supervised learning, which is the subject of the discussion in section 4.4. Kohonen self-organizing as well as Hopfield networks are presented in sections 4.5 and 4.6, respectively. Finally section 4.7 presents the conclusions of this chapter.

4.2 NN CLASSIFICATIONS

4.2.1 FEEDFORWARD AND FEEDBACK NETWORKS

In a feedforward neural network structure, the only appropriate connections are between the outputs of each layer and the inputs of the next layer. Therefore, no connections exist between the outputs of a layer and the inputs of either the same layer or previous layers. Figure 4.1 shows a two-layer feedforward network. In this topology, the inputs of each neuron are the weighted sum of the outputs from the previous layer. There are weighted connections between the outputs of each layer and the inputs of the next layer. If the weight of a branch is assigned a zero, it is equivalent to no connection between correspondence nodes. The inputs are connected to each neuron in hidden layer via their correspondence weights. Outputs of the last layer are considered the outputs of the network.

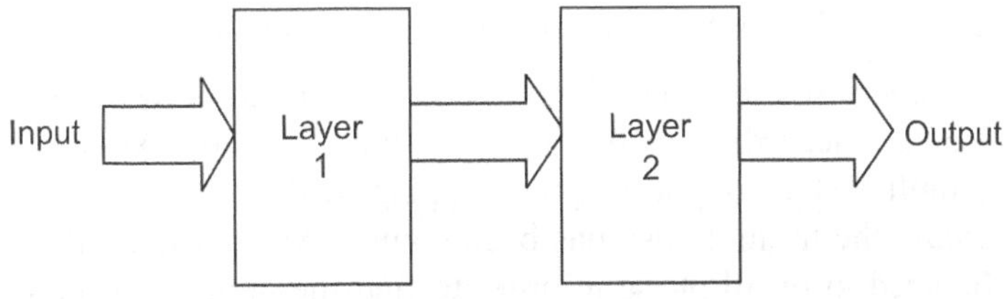

Figure 4.1 General Structure of Two-Layer Feedforward Network.

For feedback networks the inputs of each layer can be affected by the outputs from previous layers. In addition, self feedback is allowed. Figure 4.2 shows a simple single layer feedback neural network.

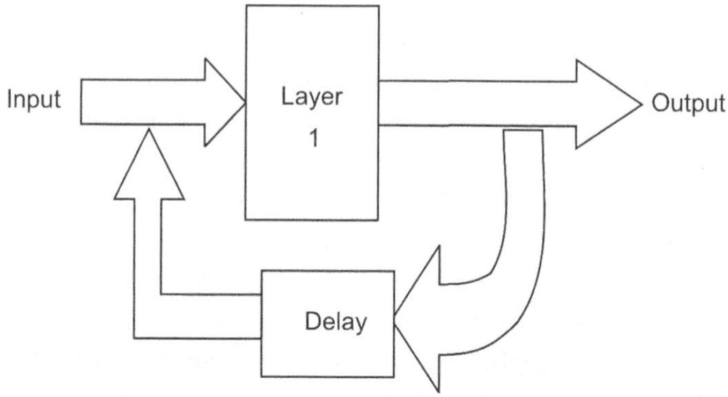

Figure 4.2 General Structure of a Sample Feedback Network.

As observed, the inputs of the network consist of both external inputs and the network output with some delays. Examples of feedback algorithms include the Hopfield network, described in detail in section 4.6, and the Boltzman Machine. An important issue for feedback networks is the stability and convergence of the network.

4.2.2 SUPERVISED AND UNSUPERVISED LEARNING NETWORKS

There are a number of approaches for training neural networks. Most fall into one of two modes:

- **Supervised Learning:** Supervised learning requires an external teacher to control the learning and incorporates global information. The teacher may be a training set of data or an observer who grades the performance. Examples of supervised learning algorithms are the least mean square (LMS) algorithm and its generalization, known as the back propagation algorithm[1]-[4], and radial basis function network [5]-[8]. They will be described in the following sections of this chapter.

  In supervised learning, the purpose of a neural network is to change its weights according to the inputs/outputs samples. After a network has established its input output mapping with a defined minimum error value, the training task has been completed. In sequel, the network can be used in recall phase in order to find the outputs for new inputs. An important factor is that the training set should be comprehensive and cover all the practical areas of applications of the network. Therefore, the proper selection of the training sets is critical to the good performance of the network.

- **Unsupervised Learning:** When there is no external teacher, the system must organize itself by internal criteria and local information designed into the network. Unsupervised learning is sometimes referred to as *self-organizing learning*, i.e., learning to classify without being taught. In this category, only the input samples are available and the network classifies the input patterns into different groups. Kohonen network is an example of unsupervised learning.

4.3 BACK PROPAGATION ALGORITHM

Back propagation algorithm is one of the most popular algorithms for training a network due to its success from both simplicity and applicability viewpoints. The algorithm consists of two phases: *Training phase and recall phase*. In the *training* phase, first, the weights of the network are randomly initialized. Then, the output of the network is calculated and compared to the desired value. In sequel, the error of the network is calculated and used to adjust the weights of the output layer. In a similar fashion, the network error is also propagated backward and used to update the weights of the previous layers. Figure 4.3 shows how the error values are generated and propagated for weights adjustments of the network.

In the *recall* phase, only the feedforward computations using assigned weights from the training phase and input patterns take place. Figure 4.3 shows both the feedforward and back propagation paths. The feedforward process is used in both recall and training phases. On the other hand, as shown in Figure 4.4(b), back propagation of error is only utilized in the training phase.

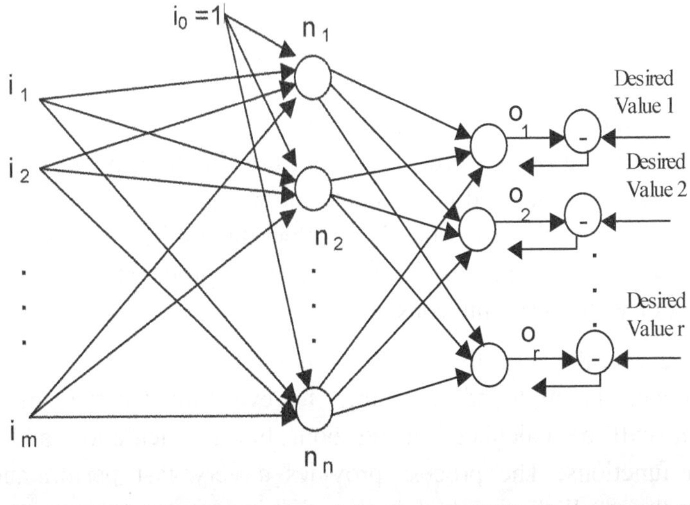

Figure 4.3 Back Propagation of the Error in a Two-Layer Network.

In the training phase, the weight matrix is first randomly initialized. In sequel, the output of each layer is calculated starting from the input layer and moving forward toward the output layer. Thereafter, the error at the output layer is calculated by comparison of actual output and the desired value to update the weights of the output and hidden layers.

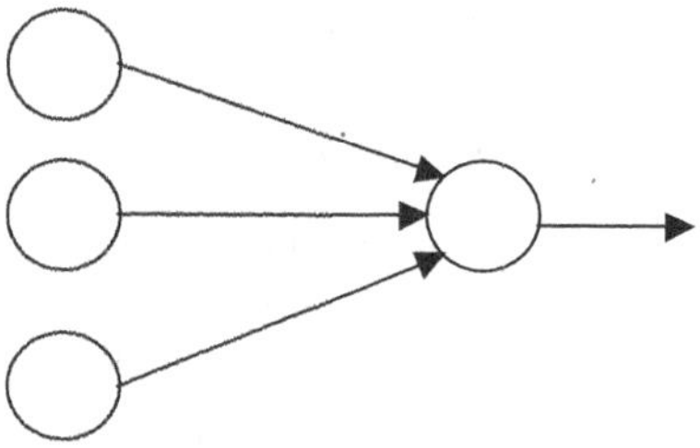

a) Forward propagation (Training and Recall Phase)

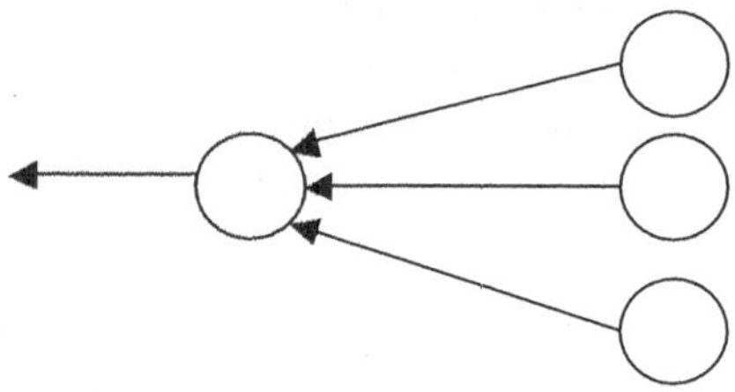

b) Backward propagation (Training Phase)

Figure 4.4 Forward Propagation in Recall and Training Phase and Backward Propagation in Training Phase.

There are two different methods of updating the weights. In the first method, weights are updated for each of the input patterns using an iteration method. In the second method, an overall error for all the input output patterns of training sets is calculated. In other words, either each of the input patterns or all of the patterns together can be used for updating the weights. The training phase will be terminated when the error value is less than the minimum set value provided by the designer. One of the disadvantages of back propagation algorithm is that the training phase is very time consuming.

During the recall phase, the network with the final weights resulting from the training process is employed. Therefore, for every input pattern in this phase, the output will be calculated using both linear calculation and nonlinear activation functions. The process provides a very fast performance of the network in the recall phase, which is one of its important advantages.

4.3.1 DELTA TRAINING RULE

The back propagation algorithm is the extension of the perceptron structure as discussed in the previous chapter with the use of multiple adaptive layers. The training of the network is based on the delta training rule method. Consider a single neuron in Figure 4.5.

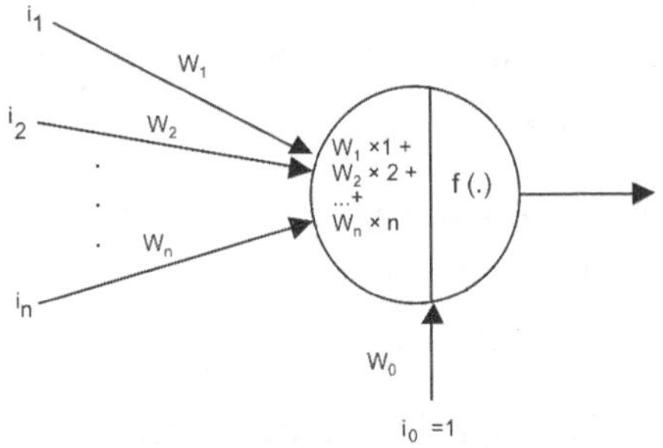

Figure 4.5 A Single Neuron.

The relations among input, activity level and output of the system can be shown as follows:

$$a = w_0 + w_1 i_1 + w_2 i_2 + \cdots + w_n i_n \quad .....(4.1)$$

or in the matrix form:

$$a = w_0 + W^T I \quad .....(4.2)$$

$$o = f(a) \quad .....(4.3)$$

where W and I are weight and input vectors of the neuron, a is activity level of the neuron and o is the output of the neuron. w_0 is called bias value. Suppose the desired value of the output is equal to d. Error e can be defined as follows:

$$e = \frac{1}{2}(d - o)^2 \quad .....(4.4)$$

by substituting Equations 4.2 and 4.4 into Equation 4.4, the following relation holds:

$$e = \frac{1}{2}(d - f(w_0 + W^T I))^2 \quad .....(4.5)$$

The error gradient vector can be calculated as follows:

$$\nabla e = -\,(d - o) f'(w_0 + W^T I) I \qquad \text{.....(4.6)}$$

The components of gradient vector are equal to:

$$\frac{\partial e}{\partial w_j} = -\,(d - o) f'(w_0 + W^T I) I_j \qquad \text{.....(4.7)}$$

where $f'(.)$ is derivative of activation function. To minimize the error, the weight changes should be in negative gradient direction. Therefore we will have

$$\Delta W = -\,\eta \nabla e \qquad \text{.....(4.8)}$$

where η is a positive constant, called learning factor. By Equations (4.6) and 4.7, the ΔW is calculated as follows:

$$\Delta W = -\eta (d - o) f'(a) I \qquad \text{.....(4.9)}$$

For each weight j, Equation 4.9 can be written as:

$$\Delta w_j = -\,\eta (d - o) f'(a) I_j \qquad j = 0, 1, 2, ..., n \qquad \text{.....(4.10)}$$

Therefore we update the weights of the network as:

$$w_j(new) = w_j(old) + \Delta w_j \qquad j = 0, 1, 2, ..., n \qquad \text{.....(4.11)}$$

For Figure 4.5, the Delta rule can be applied in a similar manner to each neuron. Through generalization of Equation 4.11 for normalized error and using Equation 4.10 for every neuron in output layer, we will have:

$$w_j(new) = w_j(old) + \frac{\eta (d_j - o_j) f'(a_j) x_j}{\|X\|^2} \qquad j = 0, 1, 2, ..., n \qquad \text{.....(4.12)}$$

where $X \in R^n$ is the input vector to the last layer, x_j is the jth element of X and $\|.\|$ denotes L2-Norm.

The above method can be applied to the hidden layers as well. The only difference is that the o_j will be replaced by y_j in 4.12. y_j is the output of hidden layer neuron, and not the output of network.

One of the drawbacks in the back propagation learning algorithm is the long duration of the training period. In order to improve the learning speed and avoid the local minima, several different methods have been suggested by researchers. These include addition of first and second moments to the learning phase, choosing proper initial conditions, and selection of an adaptive learning rate.

To avoid the local minima, a new term can be added to Equation 4.12. In such an approach, the network memorizes its previous adjustment, and, therefore it will escape the local minima, using previous updates. The new equation can be written as follows:

$$w_{j\,(new)} = w_{j\,(old)} + \frac{\eta(d_j - o_j) f'(a_j) x_j}{\|X\|^2} + \alpha[w_{j\,(new)} - w_{j\,(old)}] \qquad \text{.....(4.13)}$$

where α is a number between 0 and 1, namely the momentum coefficient.

Nguyen and Widrow have proposed a systematic approach for the proper selection of initial conditions in order to decrease the training period of the network. Another approach to improve the convergence of the network and increase the convergence speed is the adaptive learning rate. In this method, the learning rate of the network (η) is adjusted during training. In the first step, the training coefficient is selected as a large number, so the resulting error values are large. However, the error will be decreased as the training progresses, due to the decrease in the learning rate. It is similar to coarse and fine tunings in selection of a radio station.

In addition to the above learning rate and momentum terms, there are other neural network parameters that control the network's performance and prediction capability. These parameters should be chosen very carefully if we are to develop effective neural network models. Two of these parameters are described below.

Selection of Number of Hidden Layers

The number of input and output nodes corresponds to the number of network inputs and desired outputs, respectively. The choice of the number of hidden layers and the nodes in the hidden layer(s) depends on the network application. Selection of the number of hidden layers is a critical part of designing a network and is not as straightforward as input and output layers. There is no mathematical approach to obtain the optimum number of hidden layers, since such selection is generally fall into the application oriented category. However, the number of hidden layers can be chosen based on the training of the network using various configurations, and selection of the configuration with the fewest number of layers and nodes which still yield the minimum root-mean-squares (RMS) error quickly and efficiently. In general, adding a second hidden layer improves the network's prediction capability due to the nonlinear separability property of the network. However, adding an extra hidden layer commonly yields prediction capabilities similar to those of two-hidden layer networks, but requires longer training times due to the more complex structures. Although using a single hidden layer is sufficient for solving many functional approximation problems, some problems may be easier to solve with a two-hidden-layer configuration.

Normalization of Input and Output Data Sets

Neural networks require that their input and output data be normalized to have the same order of magnitude. Normalization is very critical for some applications. If the input and the output variables are not of the same order of magnitude, some variables may appear to have more significance than they actually do. The training algorithm has to compensate for order-of-magnitude differences by adjusting the network weights, which is not very effective in many of the training algorithms such as back propagation algorithm. For example, if input variable i_1 has a value of 50,000 and input variable i_2 has a value of 5, the assigned weight for the second variable entering a node of hidden layer 1 must be much greater than that for the first in order for variable 2 to have any significance. In addition, typical transfer functions, such as a sigmoid function, or a hyperbolic tangent function, cannot distinguish between two values of x_i when both are very large, because both yield identical threshold output values of 1.0.

The input and output data can be normalized in different ways. In Chapters 7 and 15, two of these normalized methods have been selected for the appropriate applications therein.

The training phase of back propagation algorithm can be summarized in the following steps:

1. Initialize the weights of the network.
2. Scale the input/output data.
3. Select the structure of the network (such as the number of hidden layers and number of neurons for each layer).
4. Choose activation functions for the neurons. These activation functions can be uniform or they can be different for different layers.
5. Select the training pair from the training set. Apply the input vector to the network input.
6. Calculate the output of the network based on the initial weights and input set.
7. Calculate the error between network output and the desired output (the target vector from the training pair).
8. Propagate error backward and adjust the weights in such a way that minimizes the error. Start from the output layer and go backward to input layer.
9. Repeat steps 5–8 for each vector in the training set until the error for the set is lower than the required minimum error.

After enough repetitions of these steps, the error between the actual outputs and target outputs should be reduced to an acceptable value, and the network is said to be trained. At this point, the network can be used in the recall or generalization phases where the weights are not changed.

Network Testing

As we mentioned before, an important aspect of developing neural networks is determining how well the network performs once training is complete. Checking the performance of a trained network involves two main criteria: (1) how well the neural network recalls the output vector from data sets used to train the network (called the *verification step*); and (2) how well the network predicts responses from data sets that were not used in the training phase (called the *recall* or *generalization step*).

In the verification step, we evaluate the network's performance in specific initial input used in training. Thus, we introduce a previously used input pattern to the trained network. The network then attempts to predict the corresponding output. If the network has been trained sufficiently, the network output will differ only slightly from the actual output data. Note that in testing the network, the weight factors are not changed: they are frozen at their last values when training ceased.

Recall or generalization testing is conducted in the same manner as verification testing; however, now the network is given input data with which it was not trained. Generalization testing is so named because it measures how well the network can generalize what it has learned, and form rules with which to make decisions about data it has not previously seen. In the generalization step, we feed new input patterns (whose results are known to us, but not to the network) to the trained network. The network generalizes well when it sensibly interpolates these new patterns. The error between the actual and predicted outputs is larger for generalization testing and verification testing. In theory, these two errors converge upon the same point corresponding to the best set of weight factors for the network.

In the following subsection, two examples are presented to clarify various issues related to BP.

Example 4.1

Consider the network (Figure 4.6) with the initial values as indicated. The desired values of the output are $d_0 = 0,\ d_1 = 1$. We show two iterations of learning of the network using back propagation. Suppose the activation function of the first layer is a sigmoid and activation function of the output is a linear function,

$$f(x) = \frac{1}{1+e^{-x}} \Rightarrow f'(x) = f(x)[1 - f(x)] \quad .....(4.14)$$

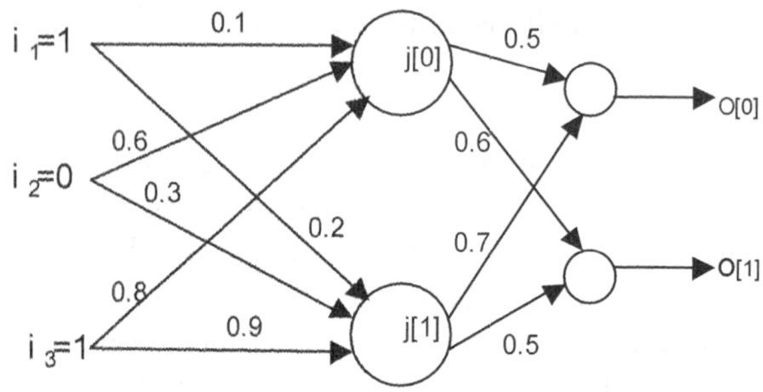

Figure 4.6 Feedforward Network of Example 4.1 with Initial Weights.

Iteration Number 1

Step 1: Initialization: First the network is initialized with the values as shown in Figure 4.6.

Step 2: Forward calculation, using Equations (4.1–4.4):

$$\text{J}[0] = f(W_j[0].I) = f(0.1\times 1 + 0.6\times 0 + 0.8\times 1 = f(0.9) = 0.7109$$
$$\text{J}[1] = f(W_j[1].I) = f(1.1) = 0.7503$$

$$\text{O}[0] = f(W_k[0]..J) = 0.5 \times 0.7109 + 0.7 \times 0.7503 = 0.88066$$
$$\text{O}[1] = f(W_k[1].J) = 0.6 \times 0.7109 + 0.5 \times 0.7503 = 0.80169$$

Step 4: According to Equation 4.5 the errors are calculated as follows:

$$\Delta k[0] = d_0 - k[0] = 0 - 0.88066 = -\ 0.88066$$
$$\Delta k[1] = d_1 - k[1] = 1 - 0.80169 = 0.19831$$

Step 4: The updated weights of the network are calculated according to Equations 4.10 and 4.11 as follows:

$$\text{W}_{\text{k00}\,(new)} = \text{W}_{\text{k00}\,(old)} + n\times \Delta k[0] \times \text{f}(\text{k}[0])\times j[0] =$$
$$0.5 + 1\times(-0.88066) + 0.2072\times 0.7109 = 0.3694$$
$$\text{W}_{\text{k01}\,(new)} = 0.56309, \qquad \text{W}_{\text{k10}\,(new)} = 0.6301, \qquad \text{W}_{\text{k11}\,(new)} = 0.5138$$
$$\text{W}_{\text{j00}\,(new)} = \text{W}_{\text{j00}\,(old)} + n\times \text{I}[0] \times \Sigma w\Delta$$
$$= 0.1 + 1\times 1\times(0.5\times -0.88066 + 0.6\times 0.19831)$$
$$= -0.2213$$

$W_{j01}(new) = 0.6;$ $W_{j02}(new) = 0.4787;$ $W_{j10}(new) = -0.3173$

$W_{j11}(new) = 0.3;$ $W_{j12}(new) = 0.3827$

Iteration Number 2: For this iteration the new weight values in Iteration 1 are utilized. Steps 2–4 of the previous iteration are repeated.

Step 2:

$J[0]=0.5640$ $J[1]=0.5163$ $O[0]=0.4991$ $O[1]=0.6299$

Step 3:

$\Delta k[0]=-0.4991$ $\Delta k[1]=0.3701$

Step 4:

$W_{k00}(new) = 0.3032$ $W_{k01}(new) = 0.5025$ $W_{k10}(new) = 0.6774$

$W_{k11}(new) = 0.5751$ $W_{j00}(new) = -0.17248$ $W_{j01}(new) = 0.6$

$W_{j02}(new) = 0.5275$ $W_{j10}(new) = -0.4015$ $W_{j11}(new) = 0.3$

$W_{j12}(new) = 0.2985$

The weights after the two iterations of training of the network can be calculated as follows:

$J[0]=0.5878;$ $J[1]=0.5257;$ $O[0]= 0.4424;$ $O[1]= 0.7005$

Table 4.1 summarizes the results for the training phase. As can be seen, the values of the output are closer to the desired value and the error value has been decreased. Training should be continued until the error values become less than a predetermined value as set by the designer (for example, 0.01). It should be noted that the selection of small values for maximum error level will not necessarily lead to better performance in the recall phase.

Table 4.1 Summary of Outputs and Error Norm after Iterations.

Error	Initial	Iteration 1	Iteration 2
Output 1	– 0.8807	– 0.4991	– 0.4424
Output 2	0.1984	0.4701	0.2995
Error Norm	0.9027	0.6214	0.5442

Choosing a very small value for this maximum error level may force the network to learn the inputs very well, but it will not lead to better overall performance.

Example 4.1 is also solved using MATLAB as shown in Chapter 21. Below is the output result of the program.

$$\begin{matrix} Final \\ Output \end{matrix} = \begin{bmatrix} -0.0088 \\ 1.0370 \end{bmatrix} \qquad \begin{matrix} Input\ \ Layer \\ Weight \end{matrix} = \begin{bmatrix} -0.0255 & 0.6 & 0.6475 \\ 0.0763 & 0.3 & 0.7763 \end{bmatrix}$$

$$\begin{matrix} Hidden\ \ Layer \\ Weight \end{matrix} = \begin{bmatrix} 0.1170 & 0.2897 \\ 0.4987 & 0.4159 \end{bmatrix} \qquad \begin{matrix} Bias \\ Weight \end{matrix} = \begin{bmatrix} -0.1255 \\ -0.1237 \end{bmatrix}$$

As observed, only four iterations are needed to complete the training task for this example. (In this case, the training sets include only one input output set, so each epoch is equivalent to an iteration.) The initial weights of the network for the program are selected as indicated in this example. The final values of the outputs are equal to -0.0088 and 1.0470. These values are close enough to the desired values. The training error is less than 0.001, which the network has achieved during the training phase.

Example 4.2

Forward Kinematics of Robot Manipulator

In this example a simple back propagation neural network has been used to solve the forward kinematic of a robot manipulator. Therefore, θ_1 θ_2 are the inputs with X, Y as the outputs of the network. A set of 200 samples is applied to the network in the training phase.

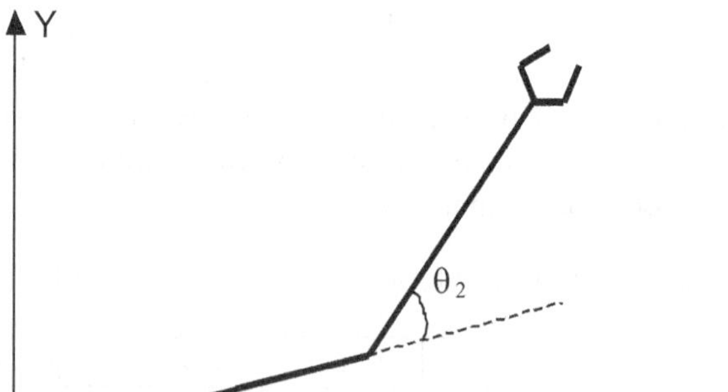

Figure 4.7 The Robot Manipulator.

The relation between (θ_1 and θ_2) and (X and Y) is as follows:

$$X = l_1 \cos\theta_1 + l_2 \cos(\theta_1 + \theta_2)$$
$$Y = l_1 \sin\theta_1 + l_2 \sin(\theta_1 + \theta_2) \qquad \text{......(4.15)}$$

Figure 4.8 shows how the error of the network changes until the performance goal has been met.

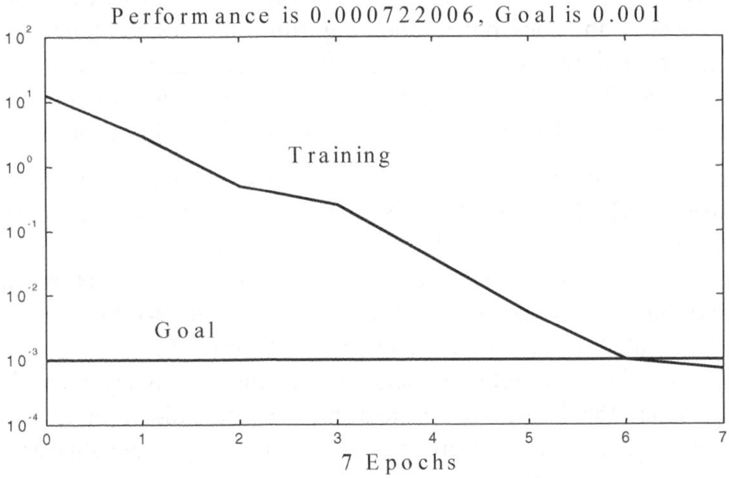

Figure 4.8 The Error of the Network During Training.

After the network has established input and output mapping during the training phase, new inputs are applied to the network to observe its performance in the recall phase. Figure 4.9 shows the simulation result of the network.

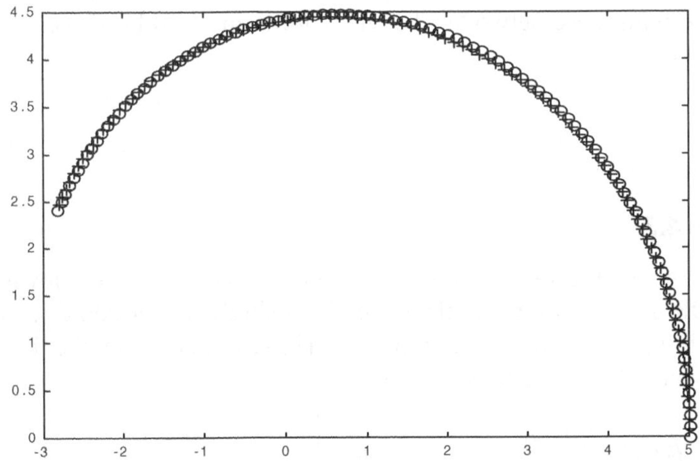

Figure 4.9 The Network Output and Prediction of the Neural Network Using the Back Propagation Algorithm.

4.4 RADIAL BASIS FUNCTION NETWORK (RBFN)

The back propagation method as described in the previous section, has been widely used to solve a number of applications [1],[2]. However, despite the practical success, the back propagation algorithm has serious training problems and suffers from slow convergence [4]. While optimization of learning rate and momentum coefficient, parameters yields overall improvements on the networks, it is still inefficient and time consuming for real time applications [4].

Radial Basis Function Networks (RBFN) provide an attractive alternative to BP networks [5]. They perform excellent approximations for curve fitting problems and can be trained easily and quickly. In addition, they exhibit none of the BP's training pathologies such as local minima problems. However, RBFN usually exhibits a slow response in the recall phase due to the large number of neurons associated in the second layer [6],[7]. One of the advantages of RBFN is the fact that linear weights associated with the output layer can be treated separately from the hidden layer neurons. As the hidden layer weights are adjusted through a nonlinear optimization, output layer weights are adjusted through linear optimization.

RBFN approximation accuracy and speed may be further improved with a strategy for selecting appropriate centers and widths of the receptive fields. The redistribution of centers to locations where input training data are meaningful can lead to more efficient RBFN [8].

In this section, the fundamental idea pertaining to the RBFN is presented. Furthermore, two examples are provided to clarify the training and recall phases associated with these networks. The network is inspired by Cover's theorem as explained below.

Cover's Theorem: A complex pattern classification problem cast in a high dimensional space nonlinearity is more likely to be linearly separable than in a low dimensional space.

Example 4.3

Consider the XOR problem as presented previously. As it was shown in chapter 2, an XOR gate cannot be implemented by a single perceptron due to nonlinear separabality property of the input pattern. However, suppose, the following pair of Guassian hidden functions are defined:

$$h_1(x) = e^{-\|x-u_1\|^2} \qquad u_1 = \begin{bmatrix} 1 \\ 1 \end{bmatrix}$$

$$h_2(x) = e^{-\|x-u_2\|^2} \qquad u_2 = \begin{bmatrix} 0 \\ 0 \end{bmatrix} \qquad \ldots\ldots(4.16)$$

If we calculate $h_1(x), h_2(x)$ for the above input patterns we will have the Table 4.2. Figure 4.10 shows the graph of the outputs in the $h_1 - h_2$ space.

Table 4.2 Mapping of XY to $h_1 - h_2$

Input pattern: X	$h_1(x)$	$h_2(x)$
(1, 1)	1	0.1454
(0, 1)	0.4678	0.4678
(0, 0)	0.1454	1
(1, 0)	0.4678	0.4678

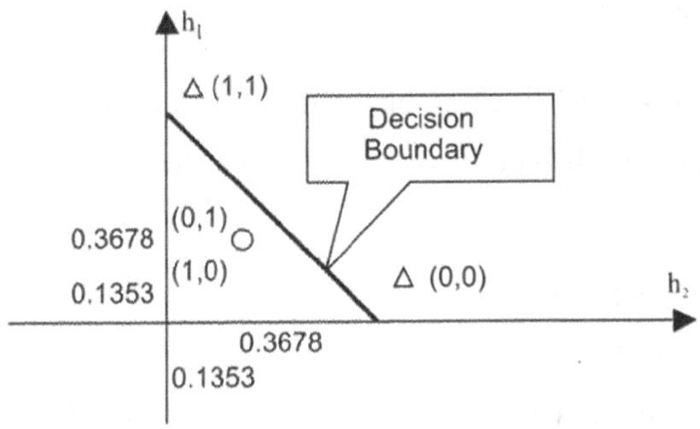

Figure 4.10 XOR Problem in $h_1 - h_2$ Space.

As can be seen, the XOR problem in $h_1 - h_2$ space is mapped to a new problem, which is linearly separable. Therefore, Guassian functions can be used to solve the above interpolation problem with one layer network. The above interpolation problem can be generalized as: Suppose there exist N points $(X_1,\ldots,X_N)$ and a corresponding set of N real values (d_1, d_2, …, d_i); find a function that satisfies the following interpolation condition:

$$F(x_i) = d_i \qquad i = 1,2,\ldots,N \qquad \ldots.(4.17)$$

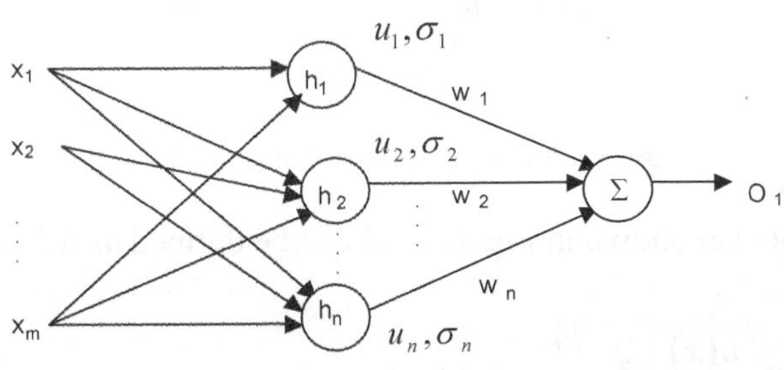

Figure 4.11 A Simple Radial Basis Network.

Figure 4.11 shows a simple radial basis network. This network is a feedforward network similar to back propagation, but it has totally different performance. The first difference is the initial weights. Despite random initial selection of the weights in back propagation, here the initial weights are not chosen randomly. The weights of each hidden layer neuron are set to values that produce a desired response. Such weights are assigned so that the network gives the maximum output for inputs equal to its weights. The activation functions h_i can be defined as follows:

$$h_i = e^{-D_i^2/2\sigma^2} \qquad \text{.....(4.18)}$$

where D_i is defined as the distance of the input to the center of the neuron which is identified by the weight vector of hidden layer neuron i. eq. (4.19) shows this relation:

$$\begin{cases} D_i^2 = (x-u_i)^T(x-u_i) \\ x: \textit{input vector} \\ u_i: \textit{Weight vector of hidden layer neuron } i \end{cases} \qquad \text{.....(4.19)}$$

Therefore, the final contribution of the neuron will decrease for the inputs, which are far from the center of the neuron. With this fact in mind, it is reasonable to give the values of each input of the training set to a neuron, which will result in faster training of the network. The main part of the training of the network is adjusting the weights of the output layer. Figure 4.12 shows a single neuron.

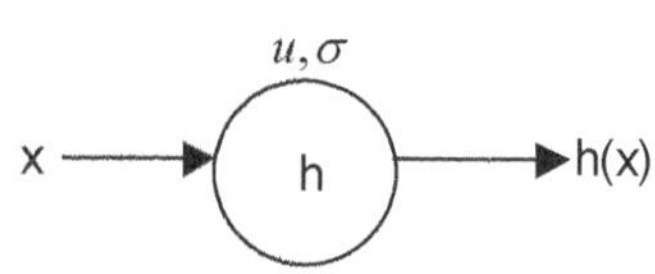

Figure 4.12 A Simple Radial Basis Neuron

Function h(x) as shown in Figure 4.13 can be defined as follows:

$$h(x) = e^{-\frac{(x-u)^2}{2\sigma^2}} \qquad \text{.....(4.20)}$$

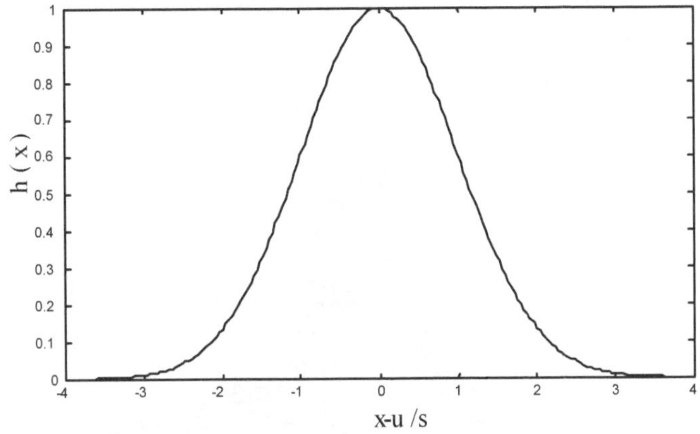

Figure 4.13 The Graph of h(x).

As both graph and formula show:

$$\begin{cases} h(x) = 1 & x = u \\ h(x) = 0 & |x-u| > 3\sigma \\ 0 < h(x) < 1 & |x-u| < 3\sigma \end{cases} \quad \text{.....(4.21)}$$

The above formula indicates that each neuron only possesses contributions from the inputs that are close to the center of the weight function. For other values of x, the neuron will have zero output value with no contribution in the final output of the network. Figure 4.14 shows a radial basis neuron with two inputs, X_1 and X_2.

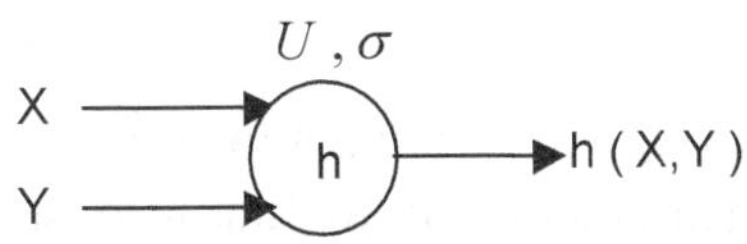

Figure 4.14 A Simple Radial Basis Neuron with Two Inputs.

Figure 4.15 shows the three-dimensional graph of this neuron. As is seen, the fundamental idea is similar. As Figure 4.15 shows, the function is radially symmetric around the center U.

Training of the radial basis network includes two stages. In the first stage, the center U_i and diameter of receptive σ_i of each neuron will be assigned. At the second stage of the training, the weight vector W will be adjusted accordingly. After the training phase is completed, the next step is the recall phase in which the outputs are applied and the actual outputs of the network are produced.

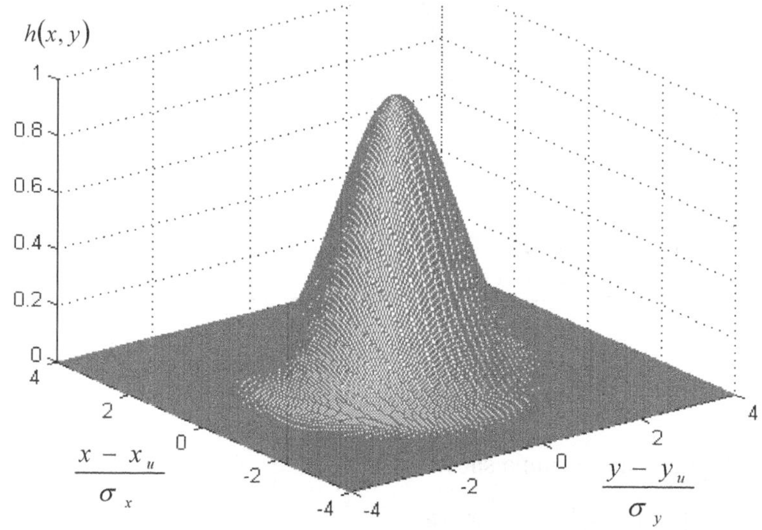

Figure 4.15 Graph of h(x,y) for the Neuron with Two Inputs.

Finding the center U_i of each neuron: One of the most popular approaches to locate the centers U_i is to divide the input vector to some clusters and then find the center of each cluster and locate a hidden layer neuron at that point.

Finding the diameter of the receptive region: The value of σ can have significant effect on the performance of the network. There are different approaches to find this value. One of the popular methods is based on the similarity of the clustering of the input data. For each hidden layer neuron, the RMS distance of each neuron and its first nearest neighbor will be calculated; this value is considered as σ.

The training phase of RBFN can be summarized as follows:

1. Apply an input vector X from the training set.
2. Calculate the output of the hidden layer.
3. Compute the output Y and compare it to the desired value. Adjust each weight W accordingly:

$$w_{ij}(n+1) = w_{ij}(n) + \eta.(x_j - y_j).x_i \qquad .....(4.22)$$

4. Repeat steps 1 to 4 for each vector in the training set.
5. Repeat steps 1 to 4 until the error is smaller than a maximum acceptable amount.

The advantage of radial basis network to back propagation network is faster training. The main problem of back propagation is its lengthy training; therefore radial basis networks have caught a lot of attention lately. The major disadvantage of radial basis network is that it is slow in the recall phase due to its nonlinear functions.

Example 4.4

This example is the same as Example 4.1, where p and o are input and output consecutively. We try to solve the problem using the radial basis network by MATLAB. The details of the program are provided in Chapter 21. The output of the program is shown below. As is observed, the output is very accurate for the same input values. Also, execution of this simple code shows that the network's training is very fast. The answer can be obtained quickly, with high accuracy. The output of the network to a similar input is also shown. $\tilde{o}$ is the output for the new applied input $\tilde{p}$, which is close to p. It can be seen that this value is close to the output of the training input.

$$p = \begin{bmatrix} 1 \\ 0 \\ 1 \end{bmatrix} \qquad o = \begin{bmatrix} 0 \\ 1 \end{bmatrix} \qquad \tilde{P} = \begin{bmatrix} 1.1 \\ -0.3 \\ 0.9 \end{bmatrix} \qquad \tilde{o} = \begin{bmatrix} 0 \\ 0.9266 \end{bmatrix}$$

Example 4.5

In this example the inverse kinematics of the robot manipulator of Example 4.2 is solved by RBFN, using MATLAB program. Figure 4.16 compares the actual path and the network prediction of this example. The actual path is shown with circles and the network output with +. As can be seen, the network can predict the path very accurately. In comparison with back propagation, prediction of RBFN is more accurate and the training of this network is much faster. However, due to the number of neurons, the recall phase of the network is usually slower than back propagation.

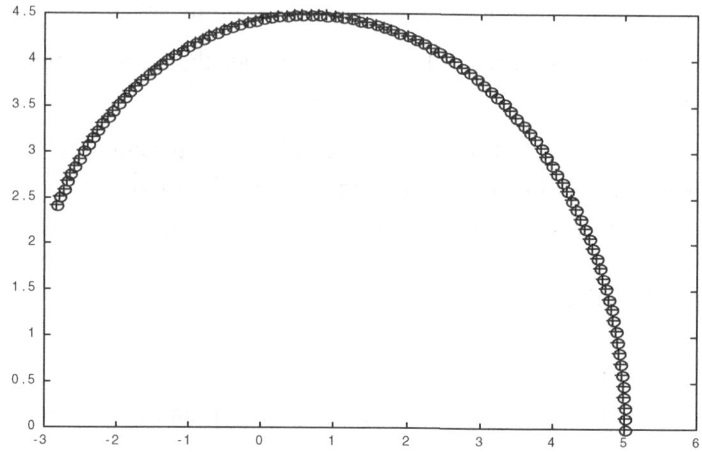

Figure 4.16 Output of the RBFN and Actual Output of the System.

4.5 KOHONEN SELF-ORGANIZATION NETWORK

The Kohonen self-organization network uses unsupervised learning and organizes itself to topological characteristics of the input patterns. The discussion in this section will not seek to explain fully all the intricacies involved in self-organization networks, but rather seek to explain the simple operation of the network with two examples. Interested readers can refer to Kohonen[10], Zurada[11], and Haykin and Simon[12] for more detailed information on unsupervised leaning and self-organization networks.

Learning and brain development phenomena of newborns are very interesting from several viewpoints. As an example, consider how a baby learns to focus its eyes. The skill is not originally present in newborns, but they generally acquire it soon after birth. The parents cannot ask their baby what to do in order to make sense of the visual stimuli impinging on the child's brain. However, it is well known that after a few days, a newborn has learned to associate sets of visual stimuli with objects or shapes. Such remarkable learning occurs naturally with little or no help and intervention from outside. As another example, a baby learns to develop a particular trajectory to move an object or grab a bottle of milk in a special manner. How can these phenomena happen?

One possible answer is provided by a self-learning system, originally proposed by Teuvo Kohonen [10]. His work provides a relatively fast and yet powerful and fascinating model of how neural networks can self-organize. In general, the term *self-organization* refers to the ability of some networks to learn without being given the correct answer for an input pattern. These networks are often closely modeled after neurobiological systems to mimic brain processing and evolution phenomena.

A Kohonen network is not a hierarchical system, but consists of a fully interconnected array of neurons. The output of each neuron is an input to all other inputs in the network including itself. Each neuron has two sets of weights: one set is utilized to calculate the sum of weighted external inputs, and another one to control the interactions between different neurons in the network. The weights on the input pattern are adjustable, while the weights between neurons are fixed.

The other two networks that have been discussed so far in this chapter (BP and RBFN) have neurons that receive input from previous layers and generate output to the next layer or the external world. However, the neurons in the network have neither input nor output to the neurons in the same layer. On the contrary, the Kohonen network receives not only the entire input pattern into the network, but also numerous inputs from the other neurons with the same layer. A block diagram of a simple Kohonen network with N neurons is shown in Figure 4.17.

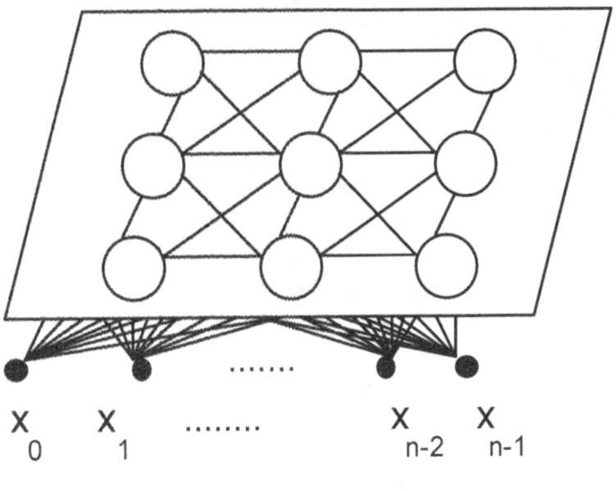

Figure 4.17 A Two Dimensional Kohonen Network.

Notice that the input is connected to all the nodes and there are interconnections between the neurons of the same layer. During each presentation, the complete input pattern is presented to each neuron. Each neuron computes its output as a sigmoidal function on the sum of its weighted inputs. The input pattern is then removed and the neurons interact with each other. The neuron with the largest activation output is declared the winner neuron and only that neuron is allowed to provide the output. However, not only the winning neuron's weight is updated, but also all the weights in a neighborhood around the winning neuron. The neighborhood size decreases slowly with each iteration [11].

4.6 HOPFIELD NETWORK

Hopfield rekindled interest in neural networks by his extensive work on different versions of the Hopfield network [14],[14]. The network can be utilized as an associative memory or to solve optimization problems. One of the original network [14], which can be used as a content addressable memory is described in this chapter. The network is a typical recursive model in which nodes are connected to one another. Figure 4.18 shows a Hopfield network.

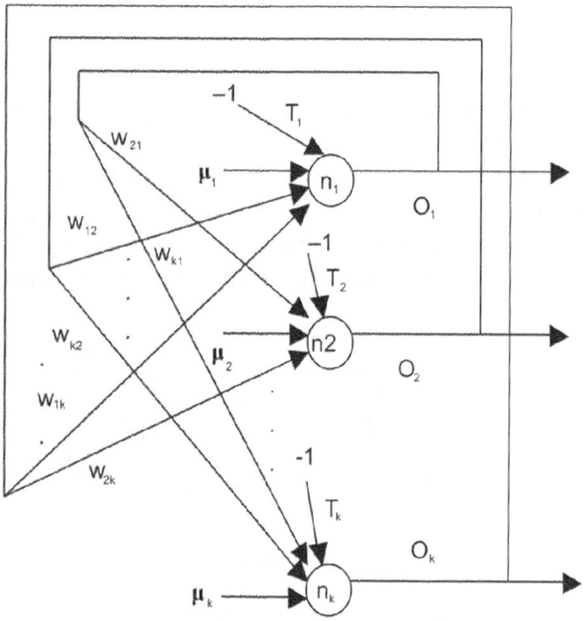

Figure 4.18 Hopfield Network.

As is shown, the output of each neuron consists of the inputs from other neurons, with the exception of itself. Therefore, the activity level of the neurons can be calculated using the following formula:

$$a_i = \sum_{\substack{j=1 \\ j \neq i}}^{n} w_{ij} o_j + \mu_i - T_i \qquad i = 1,\ 2,\ \ldots,\ n \qquad .....(4.23)$$

or in the vector form as:

$$a_i = W_i O + \mu_i - T_i \qquad i = 1,\ 2,\ \ldots,\ n \qquad .....(4.24)$$

where: $W_i = \begin{bmatrix} w_{i1} & w_{i2} & \cdots & w_{in} \end{bmatrix} \qquad i = 1,\ 2,\ \ldots,\ n$ .....(4.25)

W_i is the weight vector for the i-th input of the neural network and the i-th element of this vector is equal to zero. On the other hand,

$$O_i = \begin{bmatrix} o_1 \\ o_2 \\ \vdots \\ o_n \end{bmatrix} \qquad i = 1, \ 2, \ ..., \ n \qquad ..... (4.26)$$

is the output vector of the neural network. Equation 4.26 in the matrix form can be rewritten as follows:

$$A = WO + I - T \qquad i = 1, \ 2, \ ..., \ n \qquad .....(4.27)$$

The weight matrix W is a symmetric matrix with all diagonal elements equal to zero. If the activation function of the neuron is a sign function, we will have:

$$o_i = \begin{cases} -1 & if \quad a_i < 0 \\ +1 & if \quad a_i > 0 \end{cases} \qquad .....(4.28)$$

The output transition between old value and new value will happen at certain times. At that time, if the value of the additive weighted sum of a neuron is greater than threshold of that neuron, the new output of that neuron will remain or change to +1, otherwise it will remain or change to −1.

Considering this fact, we can define the state of the network, which is the value of the outputs at one time. For example, $O = \begin{bmatrix} 1 & -1 & 1 & \cdots & 1 \end{bmatrix}$ is a state of the network. For each neuron we have two values. Therefore 2^n states exist for a network with n neurons.

In a Hopfield network, we apply an input at certain times and then it will be removed. This causes transitions in states of the network. These transitions continue until the network reaches to a stable point, which is called an *attractor.* An important point about this network is that at each time one neuron will calculate its activity level and change its output. In other words, updating of the outputs of the neuron is being done in an asynchronous fashion. Therefore to calculate activity level of the next neuron, and find the output of that neuron, we use some updated value for the output of the other neurons. The updating order of the neurons is random. It depends on random propagation delays and noise. When using the formula in matrix form, we should be careful, because it offers synchronous or parallel updating. If we consider $E = \begin{bmatrix} 1 & -1 \end{bmatrix}$, each state of the system is an edge of the graph in E^n space. After applying an input pattern, the state of the network goes from edge to adjacent edge until it reaches an attractor of 2^n edges. An attractor should satisfy the equation:

$$\text{sgn}[A_a] = O_a \quad \text{.....(4.29)}$$

where A_a and O_a are activity level and output at the attractor. Note that if the network satisfies this equation, the next state of the network is equal to its present state and therefore no transition will happen until a new input pattern is applied to the network.

As mentioned earlier, input will be applied momentarily and then will be removed. Considering this fact and using Equation 4.27, Equation 4.29 will change to:

$$O_a = \text{sgn}[WO_a - T] \quad \text{..... (4.30)}$$

If we define the energy function for the system as:

$$\begin{aligned} E &= -\frac{1}{2} O^T W O + \mu \cdot O - T^T O \\ &= -\frac{1}{2} \sum_{i=1}^{n} \sum_{\substack{j=1 \\ j \neq i}}^{n} w_{ij} o_i o_j - \sum_{i=1}^{n} i_i o_i + \sum_{i=1}^{n} T_i o_i \end{aligned} \quad \text{..... (4.31)}$$

The gradient of the energy can be calculated from Equation 4.31 as:

$$\nabla E = -\frac{1}{2}(W^t + W)O - \mu^T + T^T = -WO - \mu^T + T^T \quad \text{.....(4.32)}$$

Here we have used the fact that the weight matrix is symmetric. The energy increment is equal to:

$$\Delta E = (\nabla E)^T \Delta O \quad \text{.....(4.33)}$$

As discussed earlier outputs will be updated one at a time. Therefore only i-th output will be updated,

$$\Delta O = \begin{bmatrix} 0 & \cdots & o_i & \cdots & 0 \end{bmatrix}^T \quad \text{.....(4.34)}$$

The energy increment will be equal to:

$$\Delta E = (-W_i^T O - \mu_i^T + T_i^T)\Delta o_i = -A_i \Delta o_i \quad \text{.....(4.35)}$$

It is obvious that for positive A_i, $\Delta o_i \geq 0$ and for negative A_i, $\Delta o_i \leq 0$. Looking at Example 4.48 it can be seen that $\Delta E \leq 0$. Therefore it can be concluded that state transitions of the network are in a way that the energy is either decreased or retained. This means that the attractors are the edges with lowest levels of energy. Following is an example to clarify these ideas.

Example 4.8

Shows the state transitions and attractors in a fourth order Hopfield network.

Consider the weight matrix as follows:

$$W = \begin{bmatrix} 0 & -1 & -1 & 2 \\ -1 & 0 & 1 & -1 \\ -1 & 1 & 0 & -1 \\ 2 & -1 & -1 & 0 \end{bmatrix} \quad .....(4.36)$$

Considering the threshold and external inputs equal to zero, energy level can be calculated as follows:

$$E = \frac{1}{2} O^T W O \quad ......(4.37)$$

or:

$$E = -\frac{1}{2} \begin{bmatrix} o_1 & o_2 & o_3 & o_4 \end{bmatrix} \begin{bmatrix} 0 & -1 & -1 & 2 \\ -1 & 0 & 1 & -1 \\ -1 & 1 & 0 & -1 \\ 2 & -1 & -1 & 0 \end{bmatrix} \begin{bmatrix} o_1 \\ o_2 \\ o_3 \\ o_4 \end{bmatrix} \quad .....(4.38)$$

After simplification we will have:

$$E = -o_1 (-o_2 - o_3 + 2o_4) - o_2 (o_3 - o_4) + o_3 o_4 \quad .....(4.39)$$

Now if we consider all the states of the network starting from $\begin{bmatrix} -1 & -1 & -1 & -1 \end{bmatrix}$ to $\begin{bmatrix} 1 & 1 & 1 & 1 \end{bmatrix}$, we can calculate all the energy levels of the network. The result will be the levels 1, 1, –1 4, –1, 4, –7, 1, 1, –7, –1, 4, 4, –1, 4, 1 respectively. Therefore the energy levels are –7, -1, 1, 4. The two states with the lowest energy level –7 are $\begin{bmatrix} -1 & 1 & 1 & -1 \end{bmatrix}$ and $\begin{bmatrix} 1 & -1 & -1 & 1 \end{bmatrix}$. We can see that these states are attractors of the network. In other words, they satisfy Example 4.44. If we try any other state of the network, we will see that they do not satisfy this equation, which means that they are not attractors of the network.

In other words, the attractors are the states with minimum levels of energy. In fact we can see that the transition in the network will be from a state to another state with a lower or the same level of energy. On the other hand, we know that the transition is asynchronous. Therefore, at each single step we will go from one state to its adjacent state. These transitions are in the direction of reduction of energy level until we reach a state with a minimum level of energy, which is the attractor of the network. Figure 4.19 shows the state transition of the network.

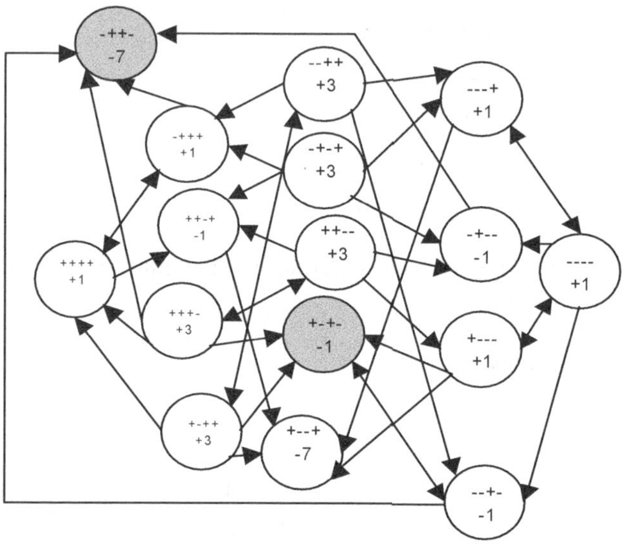

Figure 4.19 State Transition of the Hopfield Network to Reach to a Stable State.

CONCLUSIONS

The computing world has a lot to gain from neural networks. Their ability to learn by example makes them very flexible and powerful. Furthermore there is no need to devise an algorithm in order to perform a specific task; i.e. there is no need to understand the internal mechanisms of that task. They are also very well suited for real time systems because of their fast response and computational times which are due to their parallel architecture.

Neural networks also contribute to other areas of research such as neurology and psychology. They are regularly used to model parts of living organisms and to investigate the internal mechanisms of the brain.

Perhaps the most exciting aspect of neural networks is the possibility that some day 'conscious' networks might be produced. There is a number of scientists arguing that consciousness' is a 'mechanical' property and that 'conscious' neural networks are a realistic possibility.

Finally, I would like to state that even though neural networks have a huge potential we will only get the best of them when they are integrated with computing, AI, fuzzy logic and related subjects.

REFERENCES

1. Almeida L.B. (1987), "A Learning Rule for Asynchronous Perceptrons with Feedback in a Combinatorial Environment." *1st IEEE International Conference on Neural Networks*, Vol2., pp. 609-618, San Diego, CA
2. Almeida L.B. (1988), "Backpropagation in Perceptrons with Feedback", in *Neural Computers* (R. Eckmiller and C. Von der Malsburg eds.) NATO ASI Series , pp. 199-208. Newyork: Springer Verlag
3. Amari, S. (1972 a) "Learning Patterns and Pattern Sequences by Self-organizing Nets of Threshold Elements", *IEEE Trans. Computers*, C-21 (11), 1197-1206, November 1972
4. Amari, S., (1972 b), "Characteristics of Random Nets of Analog Neuron-like Elements", *IEEE Trans. on Systems, Man and Cybernetics* SMC-2, 643-657
5. Amit, D. (1989) *Modelling Brain Function,* Cambridge: Cambridge University Press. Amit, D. H. Gutfreund, and H. Sompolinsky (1985a), "Spin-Glass Models of Neural Networks", *Physical Review Letters* A 32, 1007-1018
6. Amit, D. H. Gutfreund, and H. Sompolinsky (1985b), "Storing Infinite Numbers of Patterns in a Spin-Glass Models of Neural Networks", *Physical Review Letters* 55, 1530-1533.
7. Anderson and Rosenfeld, (1988), editors, *Neurocomputing: Foundations of Research,* MIT Press, Cambridge, MA, Baddaley, A., (1983), *Your Memory,* Pelican Book, U.K
8. Bilbro G.L. and W.E Snyder (1988), "Range Image Restoration using Mean Field Annealing", *IEEE Conf. on Neural Information Processing Systems*, Denver,Nov.
9. Bressloff P.C. and D. J. Weir, (1991), "Neural Networks", *GEC Journal of Research*, Vol. 8., No 3, pp151-169
10. Broomhead D.S. and Lowe, D. (1988), "Multivariable Functional Interpolation and Adaptive Networks", Complex sYSTEMS 2, 321-355.
11. Bruck J. W. Goodman, (1988) "A generalized Convergence Theorem for Neural Networks and Its Applications in Combinatorial Optimization", *IEEE Trans. On Information Theory* 34, 1089
12. Bruck J. W. Goodman, (1990) "On the power of neural Networks for Solving Hard problems", *Journal of Complexity*, Vol. 6, pp 129-135.
13. Bullock, T. H., R. Orkand and A. Grinnell: (1977), *Introduction to Nervous System,* Freeman, San Fransisco

CHAPTER 5

ASSOCIATIVE MEMORIES

5.0 NEURAL NETWORKS AS ASSOCIATIVE MEMORY

One of the primary functions of the brain is associative memory. We associate the faces with names, letters with sounds, or we can recognize the people even if they have sunglasses or if they are somehow elder now.

Associative memories can be implemented either by using feed forward or recurrent neural networks. Such associative neural networks are used to associate one set of vectors with another set of vectors, say input and output patterns. The aim of an associative memory is, to produce the associated output pattern whenever one of the input patterns is applied to the neural network. The input pattern may be applied to the network either as input or as initial state, and the output pattern is observed at the outputs of some neurons constituting the network. According to the way that the network handles errors at the input pattern, they are classified as *interpolative* and *accretive* memory. In the interpolative memory it is allowed to have some deviation from the desired output pattern when added some noise to the related input pattern. However, in accretive memory, it is desired that the output to be exactly the same as the associated output pattern, even if the input pattern is noisy. Another classification of associative memory is such that while the memory in which the associated input and output patterns differ are called *heteroassociative memory*, it is called *autoassociative* memory if they are the same.

In this chapter, first the basic definitions about associative memory is given and then it is explained how neural networks can be made linear associators so as to perform as interpolative memory. Next it is explained how the Hopfield network can be used as autoassociative memory and then Bipolar Associative Memory network, which is designed to operate as heteroassociative memory, is introduced.

5.1 ASSOCIATIVE MEMORY

In an associative memory, we store a set of patterns μ^k, k=1...K, so that the network responds by producing whichever of the stored patterns most closely resembles the one presented to the network Here we need a measure for defining resemblance of the patterns. While Euclidean distance is convenient for the continuous valued pattern vectors, Hamming distance, which gives the number of mismatched components, is more appropriate for patterns with binary or bipolar entries. Suppose that the stored patterns, which are called exemplars or memory elements, are in the form of pairs of associations, $\mu^k=(u^k,y^k)$, $u^k.R^N$, $y^k.R^M$, k=1..K. According to the mapping.: $R^N>R^M$ that they implement, we distinguish the following types of associative memories:

- ***Interpolative associative memory:*** When u = u^r is presented to the memory, it responds by producing y^r of the stored association. However if u differs from u^r by an amount of ε, that is if u = u^r+ε is presented to the memory, then the response differs from y^r by some amount ε^r. Therefore in interpolative associative memory we have

$$\varphi(\mathbf{u}^r + \varepsilon) = \mathbf{y}^r + \varepsilon^r \text{ such that } \quad \varepsilon = 0 \Rightarrow \varepsilon^r = 0,\ k = 1..K \quad .....(5.1)$$

- ***Accretive associative memory:*** When u is presented to the memory, it responds by producing y^r of the stored association such that u^r is the one closest to u among u^k, k = *1..K*, that is,

$$\varphi(\mathbf{u} = \mathbf{y}^r) \text{ such that } \mathbf{u}^r = \min_{\mathbf{u}^k} \| \mathbf{u}^k - \mathbf{u} \|,\ k = 1..K \quad .....(5.2)$$

  The accretive associative memory in the form given above is called heteroassociative memory. However if the stored exemplars are in a special form such that the desired patterns and the input patterns are the same, that is y^k = u^k for k =*1..K*, then it is called autoassociative memory. In such a memory, whenever u is presented to the memory it responds by u^r which is the closest one to u among u^k, k = *1..K*, that is,

$$\varphi(\mathbf{u} = \mathbf{u}^r) \text{ such that } \mathbf{u}^r = \min_{\mathbf{u}^k} \| \mathbf{u}^k - \mathbf{u} \|,\ k = 1..K \quad .....(5.3)$$

  While interpolative memories can be implemented by using feed-forward neural networks, it is more appropriate to use recurrent networks as accretive memories. The advantage of using recurrent networks as associative memory is their convergence to one of a finite number of stable states when started at some initial state. The basic goals are:

- to be able to store as many exemplars as we need, each corresponding to a different stable state of the network,

- to have no other stable state
- to have the stable state that the network converges to be the one closest to the applied pattern

The problems that we are faced with being:

- The capacity of the network is restricted,
- Depending on the number and properties of the patterns to be stored, some of the exemplar may not be the stable states,
- Some spurious stable states different than the exemplars may arise by them
- The converged stable state may be other than the one closest to the applied pattern

One way of using recurrent neural networks as associative memory is to fix the external input of the network and present the input pattern u^r to the system by setting $x(0) = u^r$. If we relax such a network, then it will converge to the attractor x^*. If we are able to place each μ^k as an attractor of the network by proper choice of the connection weights, then we expect the network to relax to the attractor $x^* = \mu^r$ that is related to the initial state $x(0) = u^r$. For a good performance of the network, we need the network to converge only to one of the stored patterns μ^k, $k = 1...K$. Unfortunately, some initial states may converge to *spurious states,* which are the undesired attractors of the network representing none of the stored patterns. Spurious states may arise by themselves depending on the model used and the patterns stored. The capacity of the neural associative memories is restricted by the size of the networks. If we increment the number of stored patterns for a fixed size neural network, spurious states arise inevitably. Sometimes, the network may converge not to a spurious state, but to a memory pattern not so close to the presented pattern. What we expect for a feasible operation is, at least for the stored memory patterns themselves, if any of them is presented to the network by setting $x(0) = \mu^k$, then the network should stay converged to $x^* = \mu^r$ (Figure 5.1).

A second way to use recurrent networks as associative memory, is to present the input pattern u^r to the system as an external input. This can be done by setting $\theta = u^r$, where θ is the threshold vector whose i^{th} component is corresponding to the threshold of neuron *i*. After setting x(0) to some fixed value we relax the network and then wait until it converges to an attractor x^*. For a good performance of the network, we desire the network to have a single attractor such that $x^* = \mu^k$ for each stored input pattern u^k, therefore the network will converge to this attractor independent of the initial state of the network. Another solution to the problem is to have predetermined initial values, so that these initial values lie within the basins of attraction of μ^k whenever u^k is applied.

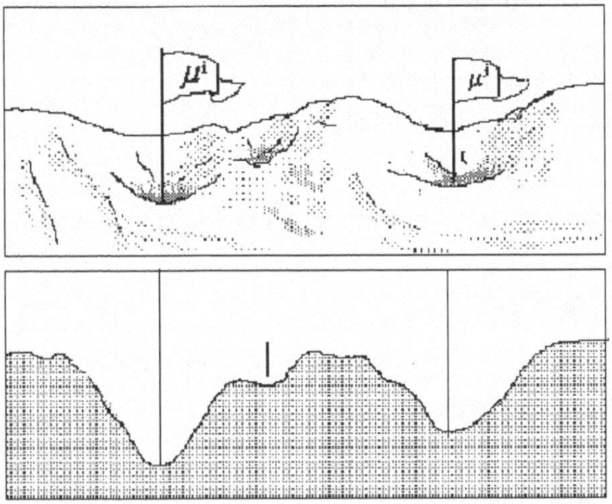

Figure 5.1 In associative memory each memory element is assigned to an attractor.

5.2 LINEAR ASSOCIATORS AS INTERPOLATIVE MEMORY

It is quite easy to implement interpolative associative memory when the set of input memory elements $\{u^k\}$ constitutes an orthonormal set of vectors, that is

$$\mathbf{u}^i.\mathbf{u}^i = \begin{cases} 1 & i = j \\ 0 & i \neq j \end{cases} \qquad .....(5.4)$$

By using kronecker delta, we write simply

$$\mathbf{u}^i.\mathbf{u}^j = \delta_{ij} \qquad .....(5.5)$$

The mapping function .(u) defined below may be used to establish an interpolative associative memory:

$$\varphi(\mathbf{u}) = \mathbf{W}^{\mathrm{T}}\mathbf{u} \qquad .....(5.6)$$

where T denotes transpose and

$$\mathbf{W} = \sum_k \mathbf{u}^k \times \mathbf{y}^k \qquad .....(5.7)$$

Here the symbol x is used to denote outer product of $x.R^N$ and $y.R^M$, which is defined as

$$\mathbf{u}^k \times \mathbf{y}^k = \mathbf{u}^k \mathbf{y}^{k^{\mathrm{T}}} = (\mathbf{y}^k \mathbf{u}^{k^{\mathrm{T}}})^{\mathrm{T}}, \qquad .....(5.8)$$

resulting in a matrix of size N by M.

By defining matrices:

$$U = [u^1\ u^2 .. u^k .. u^K] \qquad .....(5.9)$$

And

$$Y = [y^1\ y^2 .. y^k .. y^K] \qquad .....(5.10)$$

the weight matrix can be formulated as

$$W^T = YU^T \qquad .....(5.11)$$

If the network is going to be used as autoassociative memory

we have $\quad Y = U$ so,

$$W^T = UU^T \qquad .....(5.12)$$

For a function φ (u) to constitute an interpolative associative memory, it should satisfy the condition

$$\varphi(u^r) = y^r \ \textit{for } r = 1..K \qquad .....(5.13)$$

We can check it simply as

$$\varphi(u^r) = W^T u^r \qquad .....(5.14)$$

which is

$$W^T u^r = YU^T u^r \qquad .....(5.15)$$

Since the set $\{u^k\}$ is orthonormal, we have

$$\mathbf{YU}^T\mathbf{u}^r = \sum_k \delta_{kr}\mathbf{y}^k = \mathbf{y}^r \qquad .....(5.16)$$

which results in

$$\varphi(u^r) = YU^T u^r \ = y^r \qquad .....(5.17)$$

as we desired.

Furthermore, if an input pattern $u = u^r + \varepsilon$ different than the stored patterns is applied as input to the network, we obtain

$$\varphi(u) = W^T(u^r + \varepsilon)$$
$$= W^T u^r + W^T \varepsilon \qquad .....(5.18)$$

Using equation (5.2.12) and (5.2.15) results in

$$\varphi(u) = y^r + W^T \varepsilon \qquad .....(5.19)$$

Therefore, we have

$$\varphi(u) = y^r + \varepsilon^r \qquad .....(5.20)$$

in the required form, where

$$\varepsilon^r = W^T \varepsilon \qquad .....(5.21)$$

Such a memory can be implemented as shown in Figure 5.2 by using M neurons each having N inputs. The connection weights of neuron i is assigned value W_B, which is the ith column vector of matrix W. Here each neuron has a linear output transfer function $f(a)=a$. When a stored pattern u^k is applied as input to the network, the desired value y^k is observed at the output of the network as:

$$x^k = W^T u^k \quad .....(5.22)$$

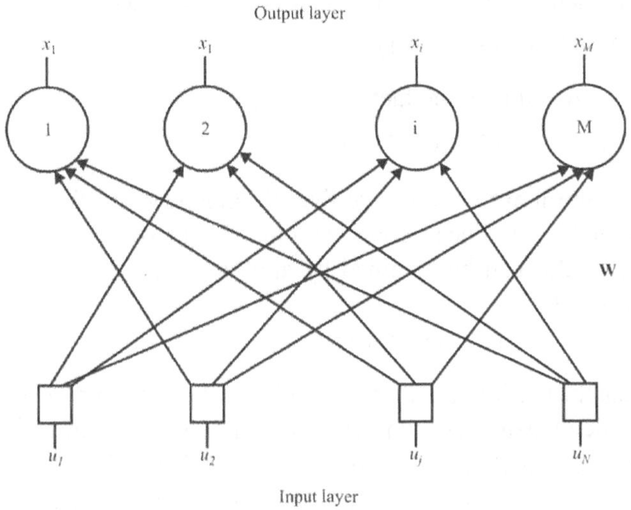

Figure 5.2 Linear Associator.

Until now, we have investigated the use of linear mapping YU^T as associative memory, which works well when the input patterns are orthonormal. In the case the input patterns are not orthonormal, the linear associator cannot map some input patterns to desired output patterns without error. In the following, we will investigate the conditions necessary to minimize the output error for the exemplar patterns. That is, for a given set of exemplars $\mu^k=(u^k,y^k)$, $u^k \in R^N$, $y^k \in R^M$, $k=1.. K$, our purpose is to find a linear mapping A* among A: $R^N \rightarrow R^M$ such that:

$$\mathbf{A}^* = \min_{\mathbf{A}} \sum_k \|\mathbf{y}^k - \mathbf{AU}^k\| \quad .....(5.23)$$

where ||.|| is chosen as Euclidean norm.

The problem may be reformulated by using the matrices U and Y as:

$$\mathbf{A}^* = \min_{\mathbf{A}} \|\mathbf{Y} - \mathbf{AU}\| \quad .....(5.24)$$

The pseudo inverse method based on least squares estimation

provides a solution for the problem in which A* is determined as:

$$A^* = Y\,U^+ \qquad .....(5.25)$$

where U^+ is pseudo inverse of U.

The pseudo inverse U^+ is a matrix satisfying the condition:

$$U^+ U = 1 \qquad .....(5.26)$$

where 1 is the identity matrix. A perfect match is obtained by

using $A^* = YU^+$, since $A^* U = YU^+ U = Y$ .....(5.27)

resulting in no error due to the fact

$$\|\mathbf{Y} - \mathbf{A}^* \mathbf{U}\| = 0 \qquad .....(5.28)$$

In the case the input patterns are linearly independent, that is none of them can be obtained as a linear combinations of the others, then a matrix U^+ satisfying eq. (5.25) can be obtained by applying the formula [Golub and Van Loan 89, Haykin 94]

$$\mathbf{U}^+ = (\mathbf{U}^T\mathbf{U})^{-1}\mathbf{U}^T \qquad .....(5.29)$$

Notice that for the input patterns, which are the columns of the matrix U, to be linearly independent, the number of columns should not be more than the number of rows, that is $K \leq N$, otherwise U^TU will be singular and no inverse will exist. The condition $K \leq N$ means that the number of entries constituting the patterns restricts the capacity of the memory. At most N patterns can be stored in such a memory. This memory can be implemented by a neural network for which $W^T = YU^+$. The desired value y^k appears at the output of the network as x^k when u^k is applied as input to the network:

$$x^k = W^T u^k \qquad .....(5.30)$$

as explained in the previous section.

Notice that for the special case of orthonormal patterns that we examined previously in this section, we have

$$U^T U = 1 \qquad .....(5.31)$$

that results in the pseudo inverse, which is in the form

$$U^+ = U^T \qquad .....(5.32)$$

and therefore

$$W^T = YU^T \qquad .....(5.33)$$

as we have derived previously.

5.3 HOPFIELD AUTO ASSOCIATIVE MEMORY

In this section we will investigate how Hopfield network can be used as auto associative memory. For this purpose some modifications are done on it so that it works in discrete state space and discrete time. When discrete Hopfield network was introduced as associative memory it had attracted a great attention. In it is shown that many important characteristics of the discrete and continuous deterministic models are closely related (Figure 5.3)

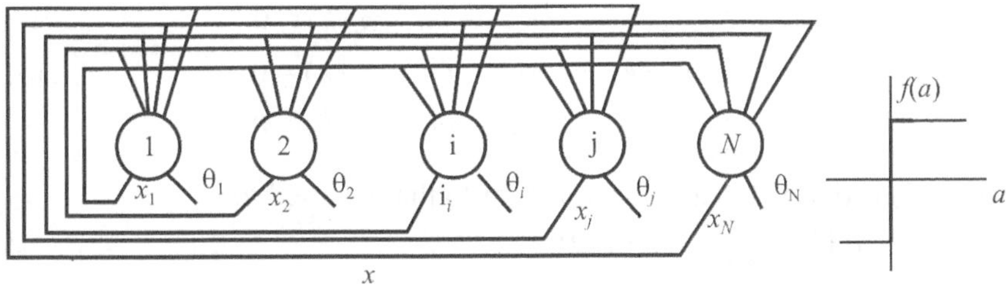

Figure 5.3 Hopfield Associative Memories.

Note that, whenever the patterns to be stored in Hopfield network are from N dimensional bipolar space constituting a hypercube, that is $u^k \in \{-1,1\}^N$, $k=1..K$, then it is convenient to have any stable state of the network on the corners of the hypercube. For this purpose refer to the output transfer function given by Eq. (2.9.5) and to Figure 2.5 for different values of the gain. If we let the output transfer function of the neurons in the network to have a very high gain,

in the extreme case

$$f_i(a) = \lim_{k \to \infty} \tanh(ka) \qquad .....(5.34)$$

$$\text{We obtain } f_i(a) = \text{sign(a)} = \begin{cases} 1 & \text{for } a > 0 \\ 0 & \text{for } a = 0 \\ -1 & \text{for } a > 0 \end{cases} \qquad .....(5.35)$$

Furthermore note that the second term of the energy function given by Eq. (2.9.5) that we repeat here for convenience:

$$E = -\frac{1}{2}\sum_{i=1}^{N}\sum_{j=1}^{N} w_{ji}x_jx_i + \sum_{i=1}^{N}\frac{1}{R_t}\int_0^{x1} f^{-1}(x)dx - \sum_{i=1}^{N}\theta_i x_i \qquad .....(5.36)$$

approaches to zero. Therefore the stable states of the network corresponds to the local minima of the function:

$$E = -\frac{1}{2}\sum_i\sum_j w_{ji}x_jx_i - \sum_i \theta_i x_i \qquad .....(5.37)$$

so that they lie on the corners of the hypercube as explained previously. However in this section we investigate a special case of the Hopfield network where the stable states of the network are forced to take discrete values in bipolar state space. Knowing in advance that the local minima of the energy function should take place at the corners of the N dimensional hypercube, we can get rid of the slow convergence problem due to small value of η.

For this purpose a discrete state excitation of the network, is provided in the following:

$$x_i(k+1) = f(a_i(k)) = \begin{cases} 1 & for \quad a_i(k) > 0 \\ x(k) & for \quad a_i(k) = 0 \\ -1 & for \quad a_i(k) > 0 \end{cases} \qquad .....(5.38)$$

where $a_i(k)$ is defined in a manner similar to that we used to:

$$a_i(k) = \sum_j w_{ji} x_j(k) + \theta_i \qquad .....(5.39)$$

The processing elements of the network are updated one at a time, such that all of the processing elements must be updated at the same average rate.

Note that, for any vector x having bipolar entries, that is $x_i \in \{-1,1\}$, we obtain the vector itself if we apply the function defined by Eq. (5.3.5) on it, that is

$$f(x) = x \qquad .....(5.40)$$

Here f is used to denote the vector function such that the function f is applied at each entry.

For stability of the discrete Hopfield network, it is further required $w_{ii} = 0$ in addition to the constraint $w_{ij} = w_{ji}$. In order to use discrete Hopfield network as autoassociative memory, its weights are fixed to

$$W^T = UU^T \qquad .....(5.41)$$

where U is the input pattern matrix as defined in Eq. (5.6). Remember that in autoassociative memory we have Y = U, where Y is the matrix of desired output patterns as defined in Eq (5.5). For the stability of the network, the diagonal entries of W is set to 0, that is $w_{ii} = 0$, $i = 1..N$.

If all the states of the network are to be updated at once, then the next state of the system can be represented in the form

$$x(k+1) = f(W^T x(k)) \qquad .....(5.42)$$

For the special case if the exemplars are orthonormal, then due to fact indicated by Eqs. (5.5) and (5.15) we have

$$f(W^T u^r) = f(u^r) = u^r \qquad .....(5.43)$$

that means each exemplar is a stable state of the network. Whenever the initial state is set to one of the exemplar, the system remains there. However, if the initial state is set to some arbitrary input, then the network converges to one of the stored exemplars, depending on the basin of attraction in which x(0) lies. However, in general, the input patterns are not orthonormal, so there is no guarantee that each exemplar is corresponding to a stable state. Therefore the problems that we mentioned in Section 5.1 arise. The capacity of the Hopfield net is less than 0.158N patterns, where N is the number of units in the network.

In the following we will show that the energy function always decreases as the state of the processing elements are changed one by one. Notice that:

$$\Delta E = E(\mathbf{X}(k+1)) - E(\mathbf{X}(k))$$

$$= -\frac{1}{2}\sum_i\sum_j w_{ji}x_j(k+1)x_i(k+1) - \sum_i \theta_i x_i(k+1) \qquad .....(5.44)$$

$$+\frac{1}{2}\sum_i\sum_j w_{ji}x_j(k)x_i(k) + \sum_i \theta_i x_i(k)$$

Assume that the neuron that just changes state at step k is neuron p. Therefore $x_p(k + 1)$ is determined by equation 5.5 and for all the other neurons we have $x_i(k+1) = x_i(k)$, $i \neq p$. Furthermore we have $w_{pp}= 0$. Hence,

$$\Delta E = -((x_p(k+1) - x_p(k))(\sum_j w_{jp}x_j(k)) + \theta_p) \qquad .....(5.45)$$

that is,

$$\Delta E = -((x_p(k+1) - x_p(k))a_p(k) \qquad .....(5.46)$$

Notice that if the value of x_p remains the same, then $x_p(k + 1) = x_p(k)$ so. E = 0. If they are not the same, than it is either the case $x_p(k) = -1$ and $x_p(k + 1) = 1$ due to fact $a_p(k) > 0$, or $x_p(k) = 1$ and $x_p(k + 1)= -1$ due to fact $a_p(k)<0$. Whatever the case is, if $x_p(k + 1) \neq x_p(k)$ it is in a direction for which $\Delta E < 0$. Therefore, for discrete Hopfield network we have

$$\Delta E \leq 0 \qquad .....(5.47)$$

Because at each state change, the energy function decreases at least by some fixed minimum amount, and because the energy function is bounded, it reaches a minimum value in a finite number of state changes. So the Hopfield network converges to one stable state in finite time in contrary to the asymptotic convergence in the continuous Hopfield network. The schedule, in which only one unit of the discrete Hopfield network is updated at a time, is called *asynchronous* update. The other approach in which all the units are updated at once is called *synchronous* update. Although the convergence with the asynchronous update mechanism is guaranteed, it may result in a cycle of length two in synchronous update.

It should be noted that, the continuous deterministic model implies the possibility of implementing the discrete network in actual hardware because of the close relation between discrete and continuous models. However, the discrete model is often implemented through computer simulations because of its simplicity.

5.4 BI-DIRECTIONAL ASSOCIATIVE MEMORY

The Bi-directional Associative Memory (BAM) is a recurrent network (Figure 5.4) designed to work as heteroassociative memory. BAM network consists of two sets of neurons whose outputs are represented by vectors x. R^N and v.R^M respectively, having activation defined by the pair of equations:

$$\frac{da_{x_t}}{dt} = -\alpha_i a_{x_f} + \sum_{j=1}^{N} w_{ji} f(a_{v_f}) + \theta_i \quad for\ i = 1.....M \qquad .....(5.48)$$

$$\frac{da_{v_f}}{dt} = -\beta_j a_{y_f} + \sum_{i=1}^{M} w_{ij} f(a_{x_t}) + \phi_j \quad for\ j = 1.....N \qquad .....(5.49)$$

where α_i, β_j, θ_i, φ_j are positive constants for all $i = 1...M, j = 1...N$,

f is sigmoid function and $W = [w_{ij}]$ is any N×M real matrix.

The stability of the BAM network can be proved easily by applying Cohen-Grossberg theorem on the state vector z. R^{N+M} defined as

$$z_i = \begin{cases} x_i & i \le M \\ v_j & j = i - M,\ M < i \le M + N \end{cases} \qquad .....(5.50)$$

that is z obtained through concatenation x and v.

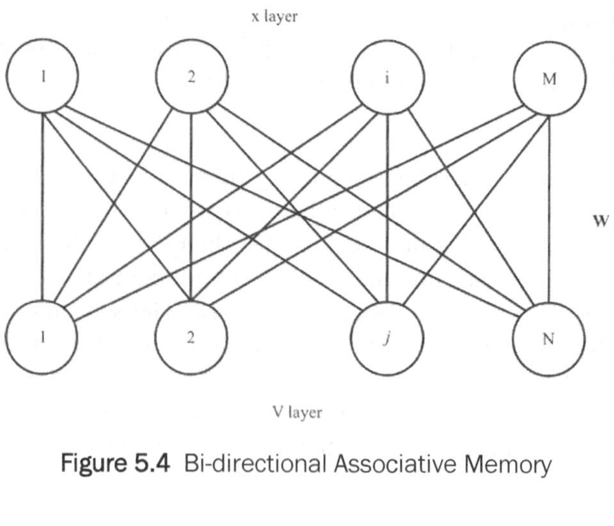

Figure 5.4 Bi-directional Associative Memory

Since BAM is a special case of the network defined by Cohen-Grossberg theorem, it has a Lyapunov Energy function as it is provided in the following:

$$E(x,v) = -\sum_{i=1}^{M}\sum_{j=1}^{N} w_{ij} f(a_{x_f}) f(a_{v_f})$$
$$+\sum_{i=1}^{M} \alpha_i \int_0^{a_{x_t}} f'(a)a \, da + \sum_{j=1}^{N} \beta_i \int_0^{a_{v_t}} f'(b)b \, db \quad .....(5.51)$$
$$-\sum_{i=1}^{M} f(x_i)\theta_i - \sum_{j=1}^{N} f(v_j)\phi_j$$

The discrete BAM model is defined in a manner similar to discrete Hopfield network. The output functions are chosen to be $f(a)$=sign(a) and states are excited as:

$$x_t(k+1) = f(a_{xt}(k)) = \begin{cases} 1 & for\ a_{xt}(k) > 0 \\ x(k) & for\ a_{xt}(k) = 0 \\ -1 & for\ a_{xt}(k) < 0 \end{cases} \quad .....(5.52)$$

$$v_f(k+1) = f(a_{v_f}(k)) = \begin{cases} 1 & for\ a_{v_f}(k) > 0 \\ v(k) & for\ a_{v_f}(k) = 0 \\ -1 & for\ a_{v_f}(k) < 0 \end{cases} \quad .....(5.53)$$

Where $a_{vf} = \sum_{i=1}^{N} w_{ij} f(a_{x_1}) + \theta_j \qquad for\ j = 1...N$ .....(5.54)

and $a_{vf} \quad \overset{N}{\underset{i\ 1}{}} \quad w_{ij} f(a_{x_1}) \quad {}_j \qquad for\ j = 1...N$ .....(5.55)

or in compact matrix notation it is shortly

$$\mathbf{X}(k+1) = \mathbf{f}(\mathbf{W}^{\mathrm{T}}\mathrm{V}(k)) \quad .....(5.56)$$

and $\mathbf{V}(\boldsymbol{k}+1) = \mathbf{f}(\mathbf{W}\mathbf{X}(\boldsymbol{k}))$ .....(5.57)

In the discrete BAM, the energy function becomes

$$\boldsymbol{E}(\mathbf{x},\mathbf{y}) = -\sum_{i=1}^{M}\sum_{j=1}^{N} w_{ij} f(a_{x_1}) f(a_{v_f})$$
$$-\sum_{i=1}^{M} f(a_{x_t})\theta_i - \sum_{j=1}^{N} f(a_{v_f})\phi_j \quad .....(5.58)$$

satisfying the condition

$$\Delta E \le 0 \qquad .....(5.59)$$

which implies the stability of the system.

The weights of BAM is determined by the equation

$$\mathbf{W}^{\mathrm{T}} = \mathbf{Y}\mathbf{U}^{\mathrm{T}} \qquad .....(5.60)$$

For the special case of orthonormal input and output patterns, we have

$$\mathbf{f}(\mathbf{W}^{\mathrm{T}}\mathbf{u}^{r}) = \mathbf{f}(\mathbf{Y}\mathbf{U}^{\mathrm{T}}\mathbf{u}^{r}) = \mathbf{f}(\mathbf{y}^{r}) = \mathbf{y}^{r} \qquad .....(5.61)$$

and

$$\mathbf{f}(\mathbf{w}\mathbf{y}^{r}) = \mathbf{f}(\mathbf{U}\mathbf{Y}^{\mathrm{T}}\mathbf{y}^{r}) = \mathbf{f}(\mathbf{u}^{r}) = \mathbf{u}^{r} \qquad .....(5.62)$$

indicating that exemplars are stable states of the network. Whenever the initial state is set to one of the exemplar, the system remains there. For arbitrary initial states the network converges to one of the stored exemplars, depending on the basin of attraction in which x(0) lies. For the input patterns that are not orthonormal, the network behaves as it is explained for the Hopfield network.

5.5 ASSOCIATIVE MEMORY NETWORKS

5.5.0 INTRODUCTORY CONCEPTS

Consider the way we are able to retrieve a pattern from a partial key as in Figure 5.5.

Figure 5.5 A key (left image) and a complete retrieved pattern (right image).

Imagine a question "what is it" in relation to the right image.

- The hood of the Volkswagen is the key to our associative memory neural network and the stored representation of the whole Volkswagen can be thought of as an network attractor for all similar keys.

- The key starts a retrieval process, which ends in an attractor, which contained both the whole car and its name (maybe you go only for the name since the question is "what is it")
- Storing a memory (an image) like the shape of a Volkswagen in an associative memory network and retrieving it, starting with a key, i.e. an incomplete version of the stored memory is the topic of this chapter.
- There are two fundamental types of the associate memory networks:
- Feed forward associative memory networks in which retrieval of a stored memory is a one-step procedure.
- Recurrent associative memory networks in which retrieval of a stored memory is a multi-step relaxation procedure. Recurrent binary associative memory networks are often referred to as the Hopfield networks.
- For simplicity we will be working mainly with binary patterns, each element of the pattern having values $\{-1, +1\}$.
- Example of a simple binary pattern:

$\xi_M =$

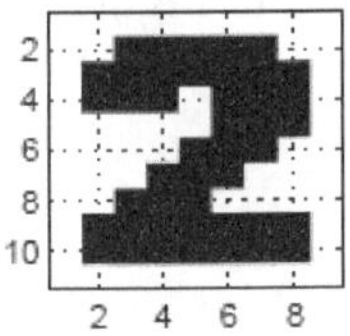

$$= \begin{bmatrix} 0&0&0&0&0&0&0&0&0 \\ 0&0&1&1&1&1&1&0&0 \\ 0&1&1&1&1&1&1&1&0 \\ 0&1&1&1&0&1&1&1&0 \\ 0&0&0&0&0&1&1&1&0 \\ 0&0&0&0&1&1&1&0&0 \\ 0&0&0&1&1&1&0&0&0 \\ 0&0&1&1&1&0&0&0&0 \\ 0&1&1&1&1&1&1&1&0 \\ 0&1&1&1&1&1&1&1&0 \\ 0&0&0&0&0&0&0&0&0 \end{bmatrix} \qquad \xi = 2\xi_M(:) - 1 = \begin{bmatrix} -1 \\ -1 \\ \vdots \\ +1 \\ +1 \\ -1 \\ \vdots \\ -1 \\ +1 \\ \vdots \\ -1 \end{bmatrix}$$

The pattern ξ is a column-scan of the matrix ξM, $\{0, 1\}$ being replaced with $\{-1, +1\}$.

5.5.1 ENCODING AND DECODING SINGLE MEMORIES

The concept of creating a memory in a neural network, that is, memorizing a pattern in synaptic weights and its subsequent retrieval is based on the "read-out" property of the outer product of two vectors.

Assume that we have a pair of column vectors:

n-component vector ξ representing the input pattern

m-component vector q representing the desired output association with the input pattern

The pair { ξ, q } to be stored is called a fundamental memory.

Encoding a single memory

We store or encode this pair in a matrix W that is calculated as an outer product (column × row) of these two vectors

$$W = q \cdot \xi^T \qquad .....(5.63)$$

Decoding a single memory

The retrieval or decoding of the store pattern is based on application of the input pattern x to the weight matrixW. The result can be calculated as follows:

$$y = W \cdot \xi = q \cdot \xi^T \cdot \xi = ||\xi|| \cdot q \qquad .....(5.64)$$

The equation says that the decoded vector y for a given input pattern ξ (the key) is proportional to the encoded vector q, the length of the input pattern ξ being the proportionality constant.

5.5.2 FEED FORWARD ASSOCIATIVE MEMORY

The above considerations give rise to a simple feed-forward associative memory known also as the linear associator. It is a well-known single layer feed-forward network with m neurons each with p synapses as illustrated in Figure 5.6.

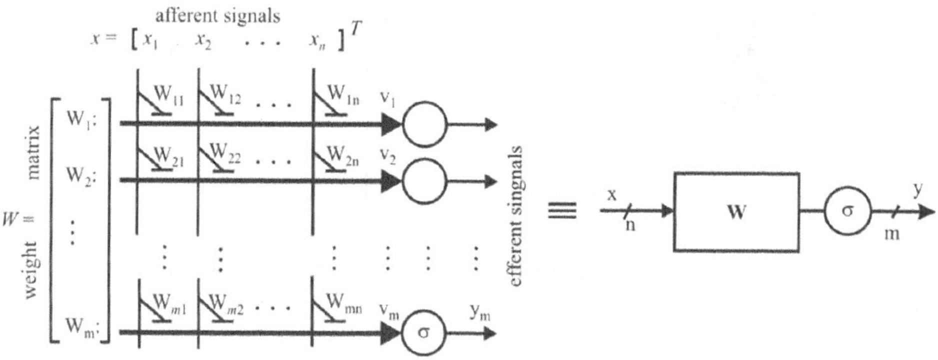

Figure 5.6 The structure of a feed-forward linear associator: y = σ (W · x)

For such a simple network to work as an associative memory, the input/output signal are binary signals with {0, 1} being mapped to {−1, +1}

- During the encoding phase the fundamental memories are stored (being encoded) in the weight matrix W
- During the decoding or retrieval phase for a given input vector x, which is the key to the memory, a specific output vector y is decoded.

5.5.3 ENCODING MULTIPLE MEMORIES

Extending the introductory concepts let us assume that we would like to store/encode K pairs of column vectors (fundamental memories) arranged in the two matrices:

$\Xi = \xi(1) \ldots \xi(K)$ a matrix of n-component vectors representing the desired input patterns

$Q = q(1) \ldots q(K)$ a matrix of m-component vectors representing the desired output associations with the input patterns

In order to encode the $\{\Xi, Q\}$ patterns we sum outer products of all pattern pairs:

$$\mathbf{W} = \frac{1}{\mathbf{K}} \sum_{k=1}^{K} q(k) \cdot \xi^{T}(k) = \frac{1}{K} Q \cdot \Xi^{T} \qquad \ldots..(5.65)$$

The sum of the outer products can be conveniently replaced by product of two matrices consisting of the pattern vectors.

The resulting $m \times n$ matrix W encodes all the desired K pattern pairs x(k), q(k).

Note that eqn (5.65) can be seen as an extension of the Hebb's learning law in which we multiply afferent and efferent signals to form the synaptic weights.

5.5.4 DECODING OPERATION

Retrieval of a pattern is equally simple and involves acting with the weight matrix on the input pattern (the key)

$$\mathbf{y} = \sigma\,(W \cdot \mathbf{x}) \qquad \ldots..(5.66)$$

where the function _ is the two-valued sign function:

$$y_j \qquad (v_j) \qquad \begin{matrix} 1 & \text{if } v_j \quad 0 \\ 1 & \text{otherwise} \end{matrix} \qquad \ldots..(5.67)$$

It is expected that

1. $x = \xi$

   If the key (input vector) x is equal to one of the fundamental memory vectors _, then the decoded pattern y will be equal to the stored/encoded pattern q for the related fundamental memory.

2. $x = \xi + n$

   If the key (input vector) x can be considered as one of the fundamental memory vectors _, corrupted by noise _ then the decoded pattern y will be also equal to the stored/encoded pattern q for the related fundamental memory.

3. $x \neq \xi + n$

   If the key (input vector) x is definitely different to any of the fundamental memory vectors ξ , then the decoded pattern y is a spurious pattern.

   - The above expectations are difficult to satisfy in a feed forward associative memory network if the number of stored patterns K is more than a fraction of m and n.
   - It means that the memory capacity of the feed forward associative memory network is low relative to the dimension of the weight matrix W.

   In general, associative memories also known as content-addressable memories (CAM) are divided in two groups:

Auto-associative: In this case the desired patterns Ξ are identical to the input patterns X, that is, $Q = \Xi$. And also n = m.

Eqn (5.65) describing encoding of the fundamental memories can be now written as:

$$W = \frac{1}{K}\sum_{k=1}^{K} \xi(k)\ .\ (k) = \frac{1}{K}\Xi\ .\ \Xi^{T}$$

Such a matrix W is also known as the auto-correlation matrix.

Hetero-associative: In this case the input Ξ and stored patterns Q and are different.

5.5.5 NUMERICAL EXAMPLES

Assume that a fundamental memory (a pattern to be encoded) is

$$\xi = \begin{bmatrix} 1 & -1 & 1 & 1 & 1 & -1 \end{bmatrix}^{T}$$

The weight matrix that encodes the memory is:

$$W = \xi\ .\ \xi^{T} = \begin{bmatrix} 1 \\ -1 \\ 1 \\ 1 \\ 1 \\ -1 \end{bmatrix} . \begin{bmatrix} 1 & -1 & 1 & 1 & 1 & -1 \end{bmatrix} = \begin{bmatrix} 1 & -1 & 1 & 1 & 1 & -1 \\ -1 & 1 & -1 & -1 & -1 & 1 \\ 1 & -1 & 1 & 1 & 1 & -1 \\ 1 & -1 & 1 & 1 & 1 & -1 \\ 1 & -1 & 1 & 1 & 1 & -1 \\ -1 & 1 & -1 & -1 & -1 & 1 \end{bmatrix}$$

Let us use the following two keys to retrieve the stored pattern:

$$W\,.\,\mathbf{x} = W\,.\,[x(1)\;X\,(2)\,] = \begin{bmatrix} 1 & -1 & 1 & 1 & 1 & -1 \\ -1 & 1 & -1 & -1 & -1 & 1 \\ 1 & -1 & 1 & 1 & 1 & -1 \\ 1 & -1 & 1 & 1 & 1 & -1 \\ 1 & -1 & 1 & 1 & 1 & -1 \\ -1 & 1 & -1 & -1 & -1 & 1 \end{bmatrix} . \begin{bmatrix} 1 & 1 \\ -1 & -1 \\ 1 & -1 \\ 1 & 1 \\ 1 & 1 \\ -1 & 1 \end{bmatrix} = \begin{bmatrix} 6 & 2 \\ -6 & -2 \\ 6 & 2 \\ 6 & 2 \\ 6 & 2 \\ -6 & -2 \end{bmatrix}$$

$$\begin{aligned} Y &= [y(1)y(2)] \\ &= \sigma(W\,.\,X) \end{aligned} \quad \left(\begin{matrix} 6 & 2 \\ 6 & 2 \\ 6 & 2 \\ 6 & 2 \\ 6 & 2 \\ 6 & 2 \end{matrix} \right) \quad \begin{matrix} 1 & 1 \\ 1 & 1 \\ 1 & 1 \\ 1 & 1 \\ 1 & 1 \\ 1 & 1 \end{matrix}$$

- The first key, x(1), is identical to the encoded fundamental memory, but the other, x(2), is different from ξ in two positions.
- However, in both cases the retrieved vectors y(1), y(2) are equal to ξ.

5.6 RECURRENT ASSOCIATIVE MEMORY— DISCRETE HOPFIELD NETWORKS

- The capacity of the feed forward associative memory is relatively low, a fraction of the number of neurons.
- When we encode many patterns often the retrieval results in a corrupted version of the fundamental memory.
- However, if we use again the corrupted pattern as a key, the next retrieved pattern is usually closer to the fundamental memory.
- This feature is exploited in the recurrent associative memory networks.

5.6.1 STRUCTURE

- A recurrent network is built in such a way that the output signals are fed back to become the network inputs at the next time step, k
- The working of the network is described by the following expressions:

$$y(k+1) = \sigma\,(W\,.\,(\mathrm{x}\,.\,\delta\,(k) + \mathrm{y}(k)) = \begin{cases} \sigma\,(W\,.\,\mathrm{x}) & \text{for } k = 0 \\ \sigma\,(W\,.\,\mathrm{y}\,(\mathrm{k})) & \textit{for } k = 1, 2, \ldots \end{cases}$$

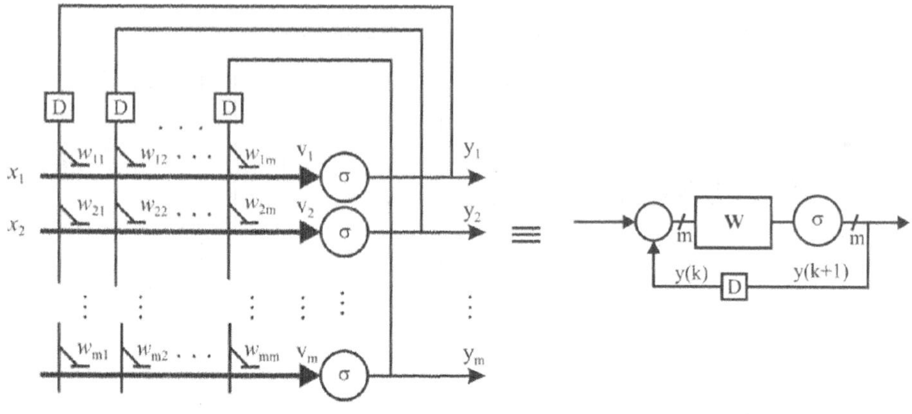

Figure 5.7 A dendritic and block diagram of a recurrent associative memory.

- The function δ(k) is called the Kronecker delta and is equal to one for k = 0, and zero otherwise. It is a convenient way of describing the initial conditions, in this case, the initial values of the input signals are equal to x(0).
- A discrete Hopfield network is a model of an associative memory, which works with binary patterns coded with {−1, +1}

Note that if v ∈ {0, 1} then u = 2v − 1 ∈ {−1, +1}

- The feedback signals y are often called the state signals.
- During the storage (encoding) phase the set of N m-dimensional fundamental memories:

  $$\Xi = [\xi\,(1),\ \xi\,(2),\ \ldots,\xi\,(K)\,]$$

  is stored in a matrix W in a way similar to the feed forward auto-associative memory networks, namely:

  $$W = \frac{1}{m}\sum_{m}^{K} \xi\,(k)\,.\,(k)^{\mathrm{T}} - \mathrm{K}\,.\,\mathrm{I}_{\mathrm{m}} = \frac{1}{m}\Xi\,.\,\Xi^{\mathrm{T}} - \mathrm{K}\,.\,\mathrm{I}_{\mathrm{m}} \qquad .....(5.68)$$

- By subtracting the appropriately scaled identity matrix I_m the diagonal terms of the weight matrix are made equal to zero, ($w_{jj} = 0$).

This is required for a stable behavior of the Hopfield network.

- During the retrieval (decoding) phase the key vector x is imposed on the network as an initial state of the network

  y(0) = x

  The network then evolves towards a stable state (also called a fixed point), such that, y(k + 1) = y(k) = y_s

It is expected that the y_s will be equal to the fundamental memory ξ closest to the key x

5.7 EXAMPLE OF THE HOPFIELD NETWORK BEHAVIOR FOR m = 3

Consider a discrete Hopfield network with three neurons as in Figure 5.8.

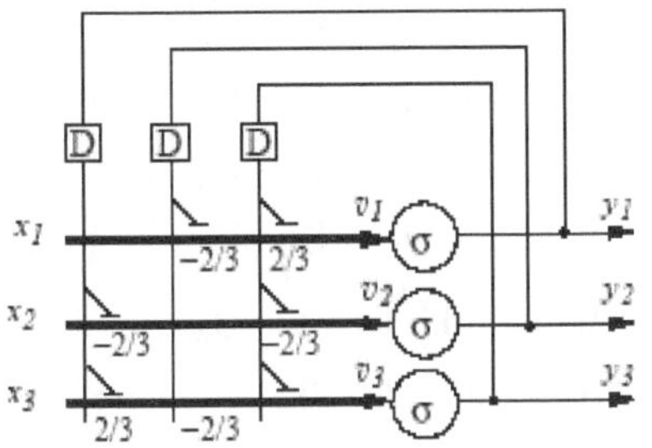

$$W = \frac{1}{3}\begin{bmatrix} 0 & -2 & +2 \\ -2 & 0 & -2 \\ +2 & -2 & 0 \end{bmatrix}$$

Figure 5.8 Example of a discrete Hopfield network with m = 3 neurons: its structure and the weight matrix

- With m = 3 binary neurons, the network can be only in $2^3 = 8$ different states.
- It can be shown that out of 8 states only two states are stable, namely: (1,−1, 1) and (−1, ,−1). In other words the network stores two fundamental memories
- Starting the retrieval with any of the eight possible states, the successive states are as depicted in Fig. 5.9.

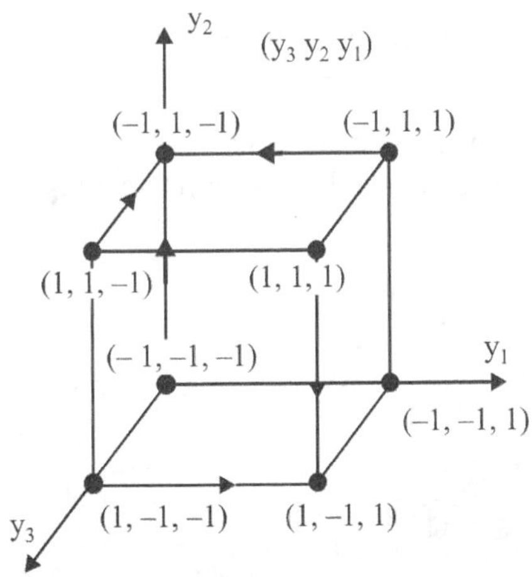

Figure 5.9 Evolution of states for two stable states.

Let us calculate the network state for all possible initial states

$$\mathbf{X} = \begin{bmatrix} y_3 \\ y_2 \\ y_1 \end{bmatrix} = \begin{bmatrix} -1 & -1 & -1 & -1 & 1 & 1 & 1 & 1 \\ -1 & -1 & 1 & 1 & -1 & -1 & 1 & 1 \\ -1 & 1 & -1 & 1 & -1 & 1 & -1 & 1 \end{bmatrix}$$

(The following MATLAB command does the trick:

X = 2*(dec2bin(0:5)–'0')'–1)

$$Y = \sigma\,(\mathrm{W}\,.\,\mathrm{X}) = \frac{1}{3}\begin{bmatrix} 0 & -2 & +2 \\ -2 & 0 & -2 \\ +2 & -2 & 0 \end{bmatrix} . \begin{bmatrix} -1 & -1 & -1 & -1 & 1 & 1 & 1 & 1 \\ -1 & -1 & 1 & 1 & -1 & -1 & 1 & 1 \\ -1 & 1 & -1 & 1 & -1 & 1 & -1 & 1 \end{bmatrix}$$

$$= \sigma\left(\frac{1}{3}\begin{bmatrix} 0 & 4 & -4 & 0 & 0 & 4 & -4 & 0 \\ 4 & 0 & 4 & 0 & 0 & -4 & 0 & -4 \\ 0 & 0 & -4 & -4 & 4 & 4 & 0 & 0 \end{bmatrix}\right) = \begin{bmatrix} 1 & 1 & -1 & 1 & 1 & 1 & -1 & 1 \\ 1 & 1 & 1 & 1 & 1 & -1 & 1 & -1 \\ 1 & 1 & -1 & -1 & 1 & 1 & 1 & 1 \end{bmatrix}$$

It is expected that after a number of relaxation steps

$$\mathrm{Y} = \mathrm{W} \cdot \mathrm{Y}$$

all patterns converge to one of two fundamental memories.

5.8 ANOTHER EXAMPLE OF HOPFIELD NETWORK (FROM LYTTON)

- The Hopfield network, or a recurrent binary associative memory consists of four neurons, each with four synapses.
- The example demonstrate the relationship between the dendritic and the flow diagram representations.
- Note that the weight matrix has the non-zero terms on the main diagonal therefore the stable pattern retrieval is not guaranteed.

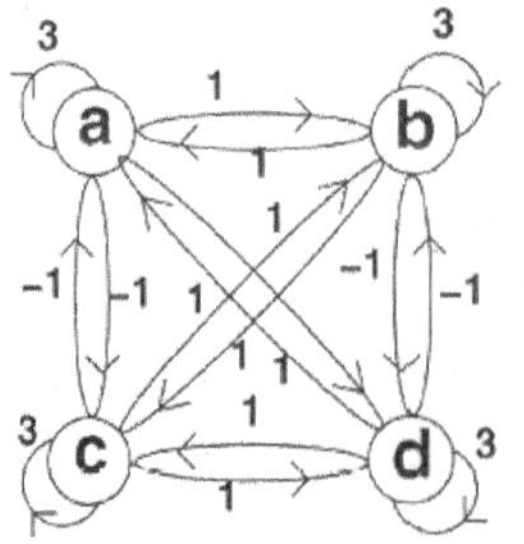

$$\mathrm{W} = \begin{bmatrix} 3 & 1 & -1 & 1 \\ 1 & 3 & 1 & -1 \\ -1 & 1 & 3 & 1 \\ 1 & -1 & 1 & 3 \end{bmatrix}$$

Stick and ball diagram of a four-unit neural networks from the summed matrix above.

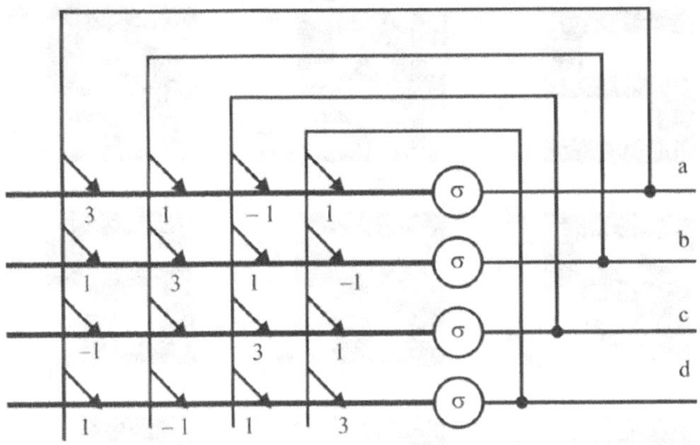

5.8.1 RETRIEVAL OF NUMERICAL PATTERNS STORED IN A RECURRENT BINARY ASSOCIATIVE MEMORY (HOPFIELD NETWORK)

The network consists of $\xi m = 120$ neurons therefore $m^2 = 14, 400$ synapses (synaptic weights) and was designed to retrieve eight digit-like patterns coded +1 for the black pixel and −1 for the white pixel (left part of the figure 5.10).

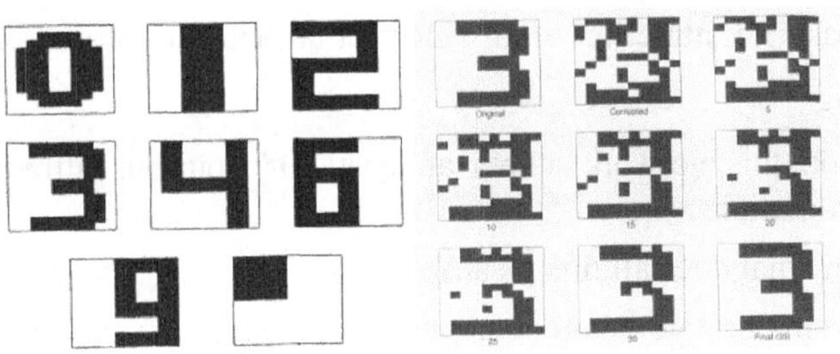

Figure 5.10 Retrieval of numerical patterns by a Hopfield network.

- To demonstrate error-correcting capability of the network, a corrupted pattern representing '3' was applied to the network. After 35 iterations the output pattern was the perfect re-call of the pattern.
- Retrieval from a corrupted pattern (key) succeeds because it was within the basin of attraction of the correct attractor, that is, a stored pattern, or fundamental memory.

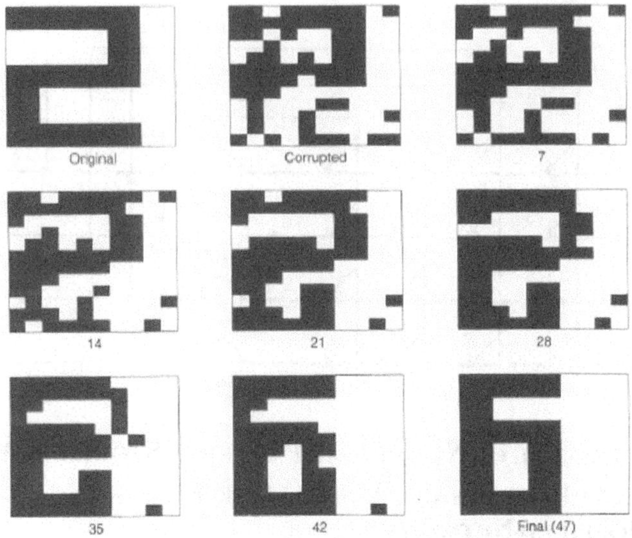

Figure 5.11 Unsuccessful retrieval of a numerical pattern by Hopfield network.

This time retrieval from a corrupted pattern does not succeed because it was within the basin of attraction of an incorrect attractor, or stored pattern. This is not surprising.

In addition this network stores at least 108 spurious attractors found in computer simulations:

- Then what does all this mean?
- It means that you cannot store memories that are similar to each other, because if you have a slightly corrupted version of one of two similar memories, then you can easily end up in the other one.
- It also means that even if the stored memories are not similar to each other, there will be other, spurious memories "in-between".
- And if your corrupted initial input is closer to its own fundamental memory than to all the other fundamental memories it is still not certain that the proper fundamental memory will be retrieved, it might well be a spurious attractor instead.

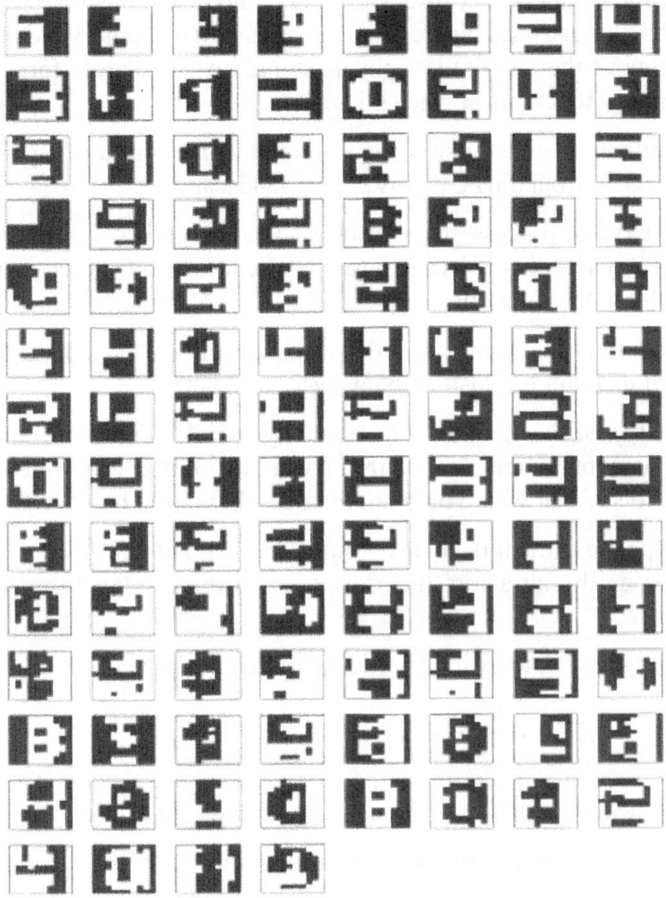

Figure 5.12 Compilation of the spurious states produced in the computer experiment on the Hopfield network

- The risk of retrieving the opposite of a fundamental memory is usually not great — your initial input has to be very corrupted for that to happen.
- All this means that the associative memory networks as we have described them are far from ideal from a legal witness point of view.
- But their shortcomings are not unheard of from human experience.
- So their shortcomings do not rule them out as first-order models of human memory.

5.9 THE ENERGY LANDSCAPE

- There is an "energy" associated with the states of a recurrent associative memory network.

- It is called energy because Hopfield, who used the energy concept to describe the retrieval process in an associative memory network (in the beginning of the 1980's) is a physicist and saw the purely formal similarity with energy functions in mechanics.
- Each attractor gives rise to a minimum, i.e., a lower point than its immediate surroundings, in this energy landscape.
- If the retrieval process starts from a corrupted memory, then it starts at a high energy and, like a ball in a real landscape, it rolls down to a minimum, hopefully to the right one.
- A problem is that the ball rolls in a high-dimensional landscape, which makes it difficult to illustrate on paper.
- The energy of the spurious attractors is generally higher than the energy of the fundamental memories, so if you can "feel this energy" then you have a chance to say that what you seem to remember might be wrong.
- The opposite attractors of the fundamental memories have the same low energies as the attractors themselves so in this case you are left without assistance.

The energy associated with a particular state y is defined as:

$$E = -\frac{1}{2}\sum_{i=1}^{m}\sum_{j=1}^{m} w_{ji} y_i y_i = -\frac{1}{2} y^{\mathrm{T}} . W . y \quad (w_{ii} = 0)$$

where m is the number of neurons, each with m synapses.

The minus sign ensures that we have minima for the "ball to roll into", rather than peaks to climb. The main diagonal terms in the weight matrix should be zero to ensure that a stable solution can be attained.

Example of an imaginary energy landscape (from Lytton):

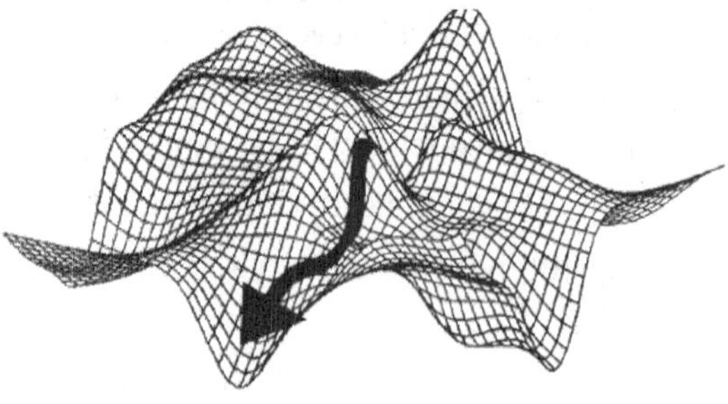

Figure 5.13 A made up memory landscape for a two dimensional network.

5.10 EXAMPLE OF A TWO-NEURON RECURRENT ASSOCIATIVE MEMORY

$$\begin{bmatrix} -1 & -1 & +1 & +1 \\ -1 & +1 & -1 & +1 \end{bmatrix}$$

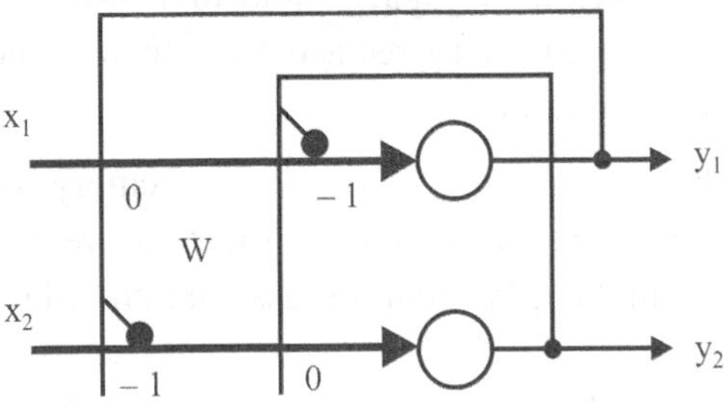

$$W = \begin{bmatrix} 0 & -1 \\ -1 & 0 \end{bmatrix}$$

The four possible states of a two-dimensional memory network are shown.

One is a fundamental memory, one is its opposite and two lie on a limit cycle as Lytton sees it.

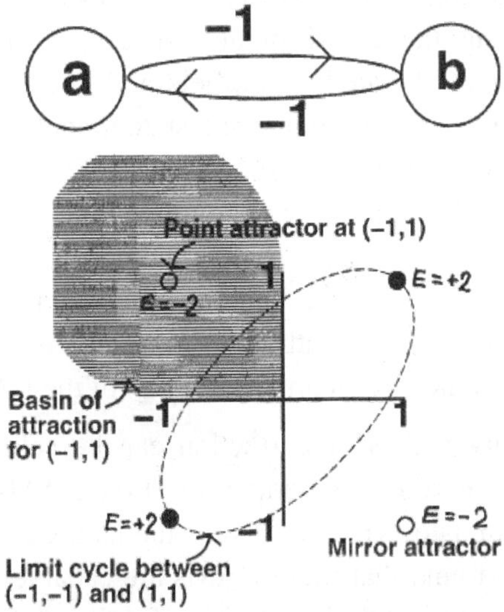

Figure 5.14 A simple two-unit mutual inhibition network and the corresponding state space diagram.

CONCLUSIONS

This chapter introduces a variety of associative neural memories and characterizes their capacity and their error correction capability. In particular, attention is given to recurrent associative nets with dynamic recollection of stored information.

The most simple associative memory is the linear associative memory (LAM) with correlation-recording of real-valued memory patterns. Perfect storage in the LAM requires associations whose key patterns (input patterns) are orthonormal. Furthermore, one only needs to have linearly independent key patterns if the projection recording technique is used. This results in an optimal linear associative memory (OLAM) which has noise suppression capabilities. If the stored associations are binary patterns and if a clipping nonlinearity is used at the output of the LAM, then the orthonormal requirement on the key patterns may be relaxed to a pseudo-orthogonal requirement. In this case, the associative memory is nonlinear.

Methods for improving the performance of LAMs, such as multiple training and adding specialized associations to the training set, are also discussed. The remainder of the chapter deals with DAMs (mainly single-layer autoassociative DAM's) which have recurrent architectures.

The stability, capacity, and associative retrieval properties of DAMs are characterized. Among the DAM models discussed are the continuous-time continuous-state model (the analog Hopfield net), the discrete-time continuous-state model, and the discrete-time discrete-state model (Hopfield's discrete net). The stability of these DAMs is shown by defining appropriate Liapunov (energy) functions. A serious shortcoming with the correlation-recorded versions of these DAMs is their inefficient memory storage capacity, especially when error correction is required. Another disadvantage of these DAMs is the presence of too many spurious attractors (or false memories) whose number grow exponentially in the size (number of units) of the DAM.

Improved capacity and error correction can be achieved in DAMs which employ projection recording. Several projection DAMs are discussed which differ in their state update dynamics and/or the nature of their state: continuous versus discrete. It is found that these DAMs are capable of storing a number of memories which can approach the number of units in the DAM. These DAMs also have good error correction capabilities. Here, the presence of self-coupling (diagonal-weights) is generally found to have a negative effect on DAM performance; substantial improvements in capacity and error correction capability are achieved when self-coupling is eliminated.

REFERENCES

1. Bounds, D. G., Lloyd, P. J., Mathew, B., and Wadell, G. (1988). "A Multilayer Perceptron Network for the Diagnosis of Low Back Pain," in *Proc. IEEE International Conference on Neural Networks* (San Diego 1988), vol. II, 481-489
2. Bourlard, H. and Kamp, Y. (1988). "Auto-Association by Multilayer Perceptrons and Singular Value Decomposition," *Biological Cybernetics*, 59, 291-294.
3. van den Bout, D. E. and Miller, T. K. (1988). "A Traveling Salesman Objective Function that Works," in *IEEE International Conference on Neural Networks* (San Diego 1988), vol. II, 299-303. IEEE, New York.
4. van den Bout, D. E. and Miller, T. K. (1989). "Improving the Performance of the Hopfield-Tank Neural Network Through Normalization and Annealing," *Biological Cybernetics*, 62, 129-139.
5. Bromley, J. and Denker, J. S. (1993). "Improving Rejection Performance on Handwritten Digits by Training with 'Rubbish'," *Neural Computation*, 5(3), 367-370.
6. Broomhead, D. S. and Lowe, D. (1988). "Multivariate Functional Interpolation and Adaptive Networks," *Complex Systems*, 2, 321-355.
7. Brown, R. R. (1959). "A Generalized Computer Procedure for the Design of Optimum Systems: Parts I and II," *AIEE Transactions, Part I: Communications and Electronics*, 78, 285-293.
8. Brown, R. J. (1964). *Adaptive Multiple-Output Threshold Systems and Their Storage Capacities*, Ph.D. Thesis, Tech. Report 6771-1, Stanford Electron. Labs, Stanford University, CA.

CHAPTER 6

INTRODUCTION TO FUZZY SETS: BASIC DEFINITIONS AND RELATIONS

6.1 INTRODUCTION

One of the more popular new technologies is "intelligent control," which is defined as a combination of control theory, operations research, and artificial intelligence (AI). Judging by the billions of dollars worth of sales and close to 2000 patents issued in Japan alone since the announcement of the first fuzzy chips in 1987, fuzzy logic still is perhaps the most popular area in AI. Thanks to tremendous technological and commercial advances in fuzzy logic in Japan and other nations, today fuzzy logic continues to enjoy an unprecedented popularity in the technological and engineering fields including manufacturing. Fuzzy logic technology is being used in numerous consumer and electronic products and systems, even in the stock market and medical diagnostics. The most important issue facing many industrialized nations in the next several decades will be global competition to an extent that has never before been posed. The arms race is diminishing and the economic race is in full swing. Fuzzy logic is but one such front for global technological, economical, and manufacturing competition.

In order to understand fuzzy logic it is important to discuss fuzzy sets. In 1965, Zadeh [1] wrote a seminal paper in which he introduced fuzzy sets, i.e., sets with unsharp boundaries. These sets are generally in better agreement with the human mind that works with shades of gray, rather than with just black or white. Fuzzy sets are typically able to represent linguistic terms, e.g., warm, hot, high, low. Nearly ten years later Mamdani [2] succeeded in applying fuzzy logic for control in practice. Today, in Japan, U.S.A, Europe, Asia and many other parts of the world fuzzy control is widely accepted and applied. In many consumer products like washing machines and cameras, fuzzy controllers are used in order to obtain intelligent machines (Intelligent Machine Quotient--MIQ®) and user friendly products. A few interesting applications can be mentioned: control of subway systems, image stabilization of video cameras, image enhancement and autonomous control of helicopters. Although the U.S and Europe hesitated in accepting fuzzy logic, they have become more enthusiastic about applying this technology.

Fuzzy set theory is developed comparing the precepts and operations of fuzzy sets with those of classical set theory. Fuzzy sets will be seen to contain the vast majority of the definitions, precepts, and axioms that define classical sets. In fact, very few differences exist between the two set theories. Fuzzy set theory is actually a fundamentally broader theory than current classical set theory, in that it considers an infinite number of "degrees of membership" in a set other than the canonical values of 0 and 1 apparent in classical set theory. In this sense, one could argue that classical sets are a limited form of fuzzy sets. Hence, it will be shown that fuzzy set theory is a comprehensive set theory.

Conceptually, a fuzzy set can be defined as a collection of elements in a universe of information where the boundary of the set contained in the universe is ambiguous, vague, and otherwise fuzzy. It is instructive to introduce fuzzy sets by first reviewing the elements of classical (crisp) set theory.

This chapter is organized as follows. Section 6.2 briefly describes classical sets, followed by introduction to classical set operations in section 6.3. Properties of classical sets are given in section 6.4. Section 6.5 is a quick introduction to fuzzy sets. Fuzzy set operations and properties are given in sections 6.6 and 6.7, respectively. Section 6.8 presents fuzzy vs. classical relations. Finally, a conclusion is given in section 6.9.

Classical Sets

In classical set theory, a set is denoted as a so-called *crisp set* and can be described by its characteristic function as follows:

$$\mu_C : U \rightarrow \{0,1\} \qquad \text{.....(6.1)}$$

In Equation 6.1, U is called the universe of discourse, i.e., a collection of elements that can be continuous or discrete. In a crisp set each element of the universe of discourse either belongs to the crisp set ($\mu_C = 1$) or does not belong to the crisp set ($\mu_C = 0$).

Consider a characteristic function μ_{Chot} representing the crisp set hot, a set with all "hot" temperatures. Figure 6.1 graphically describes this crisp set, considering temperatures higher than 40°C as hot. (Note that for all temperatures T, we have $T \in U$).

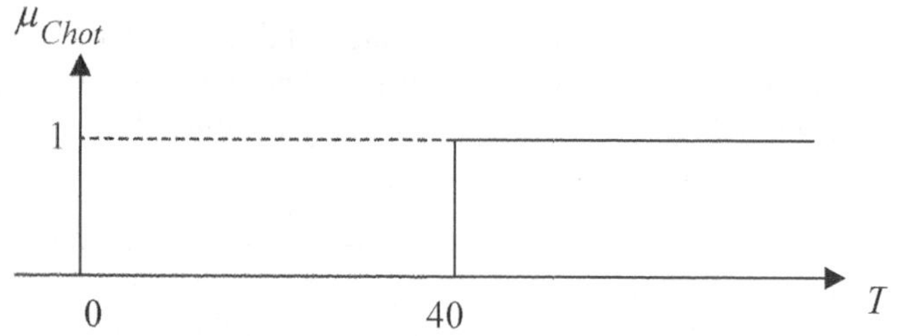

Figure 6.1 The Characteristic Function μ_{Chot}.

Classical Set Operations

Let A and B be two sets in the universe U, and $\mu_A(x)$ and $\mu_B(x)$ be the characteristic functions of A and B in the universe of discourse in sets A and B, respectively. The characteristic function $\mu_A(x)$ is defined as follows:

$$\mu_A(x) = \begin{cases} 1, & x \in A \\ 0, & x \notin A \end{cases} \qquad .....(6.2)$$

and $\mu_B(x)$ is defined as

$$\mu_B(x) = \begin{cases} 1, & x \in B \\ 0, & x \notin B \end{cases} \qquad .....(6.3)$$

Using the above definitions, the following operations are defined [3].

Union: The union between two sets, i.e., $C = A \cup B$, where $\cup$ is the union operator, represents all those elements in the universe which reside in either the set A or set B or both [4], (see Figure 6.2). The characteristic function μ_C is defined in Equation 6.4.

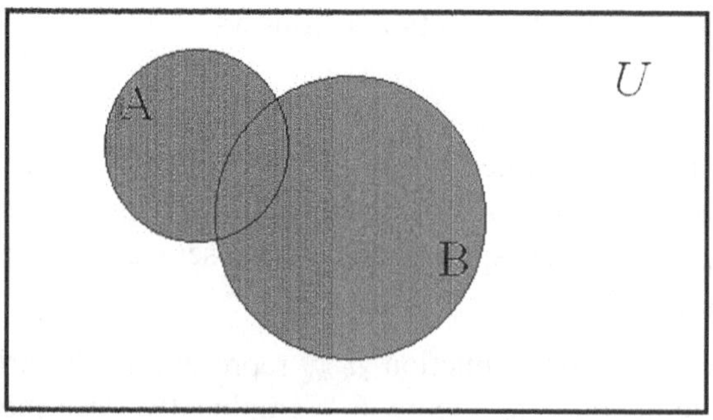

Figure 6.2 Union.

$$\forall x \in U : \mu_C = \max[\mu_A(x), \mu_B(x)] \qquad .....(6.4)$$

The operator in Equation 6.4 is referred to as the *max-operator.*

Intersection: The intersection of two sets, i.e., $C = A \cap B$, where $\cap$ is the intersection operator, represents all those elements in the universe U which reside in both sets A and B simultaneously (see Figure 6.3). Equation 6.5 shows how to obtain the characteristic function μ_C.

$$\forall x \in U : \mu_C = \min\left[\mu_A(x), \mu_B(x)\right] \qquad .....(6.5)$$

The operator in Equation 6.5 is referred to as the *min-operator.*

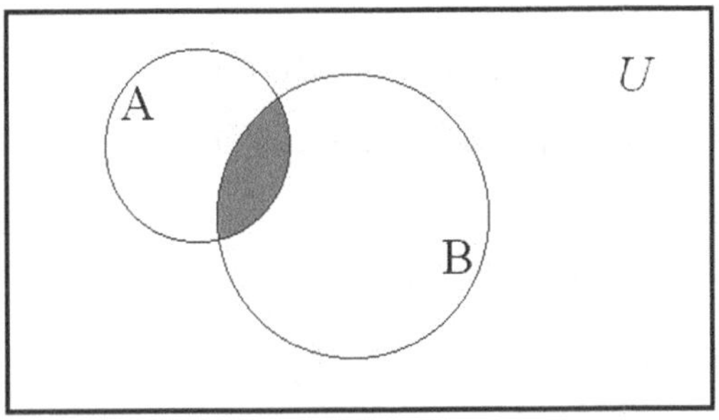

Figure 6.3 Intersection.

Complement: The complement of a set A, denoted $\overline{A}$, is defined as the collection of all elements in the universe which do not reside in the set A (see Figure 6.4). The characteristic function $\mu_{\overline{A}}$ is defined by Equation 6.6.

$$\forall x \in U : \mu_{\overline{A}} = 1 - \mu_A(x) \qquad .....(6.6)$$

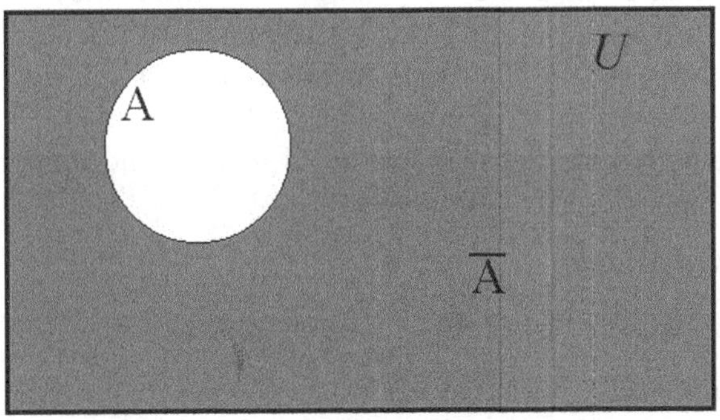

Figure 6.4 Complement.

Properties of Classical Set

Properties of classical sets are very important to consider because of their influence on the mathematical manipulation. Some of these properties are listed below [5].

Commutativity:

$$A \cup B = B \cup A \quad .....(6.7)$$

$$A \cap B = B \cap A \quad .....(6.8)$$

Associativity:

$$A \cup (B \cup C) = (A \cup B) \cup C \quad .....(6.9)$$

$$A \cap (B \cap C) = (A \cap B) \cap C \quad .....(6.10)$$

Distributivity:

$$A \cup (B \cap C) = (A \cup B) \cap (A \cup C) \quad .....(6.11)$$

$$A \cap (B \cup C) = (A \cap B) \cup (A \cap C) \quad .....(6.12)$$

Idempotency:

$$A \cup A = A \quad .....(6.13)$$

$$A \cap A = A \quad .....(6.14)$$

Identity:

$$A \cup \phi = A \quad .....(6.15)$$

$$A \cap X = A \quad .....(6.16)$$

$$A \cap \phi = \phi \quad .....(6.17)$$

$$A \cup X = X \quad .....(6.18)$$

Excluded middle laws are very important since they are the only set operations that are not valid for both classical and fuzzy sets. Excluded middle laws consist of two laws. The first, known as *Law of Excluded Middle*, deals with the union of a set A and its complement. The second law, known as *Law of Contradiction*, represents the intersection of a set A and its complement. The following equations describe these laws:

Law of Excluded Middle

$$A \cup \bar{A} = X \quad .....(6.19)$$

Law of Contradiction

$$A \cap \bar{A} = \phi \quad .....(6.20)$$

Fuzzy Sets

The definition of a fuzzy set [1] is given by the characteristic function

$$\mu_{\underset{\sim}{F}} : U \rightarrow [0,1] \quad .....(6.21)$$

In this case the elements of the universe of discourse can belong to the fuzzy set with any value between 0 and 1. This value is called the *degree of membership*. If an element has a value close to 1, the degree of membership, or truth value is high. The characteristic function of a fuzzy set is called the *membership function*, for it gives the degree of membership for each element of the universe of discourse. If now the characteristic function μ_{Fhot} is considered, one can express the human opinion, for example, that 37°C is still fairly hot, and that 38°C is hot, but not as hot as 40°C and higher. This result in a gradual transition from membership (completely true) to non-membership (not true at all). Figure 6.5 shows the membership function μ_{Fhot} for the fuzzy set F_{hot}.

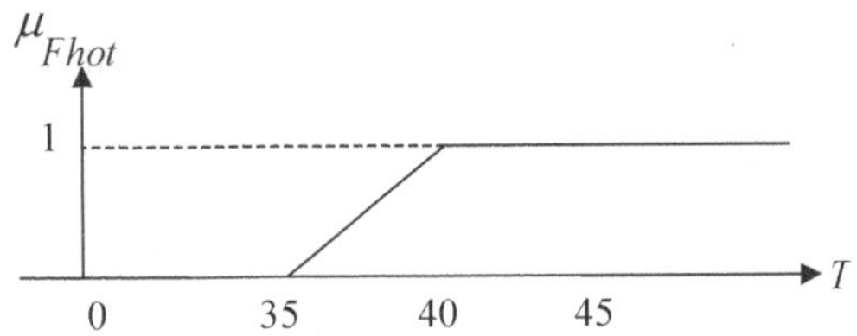

Figure 6.5 The Membership Function μ_{Fhot}..

6.1.1 FUZZY MEMBERSHIP FUNCTIONS

The membership functions for fuzzy sets can have many different shapes, depending on definition. Figure 6.6 provides a description of the various features of membership functions. Some of the possible membership functions are shown in Figure 6.7.

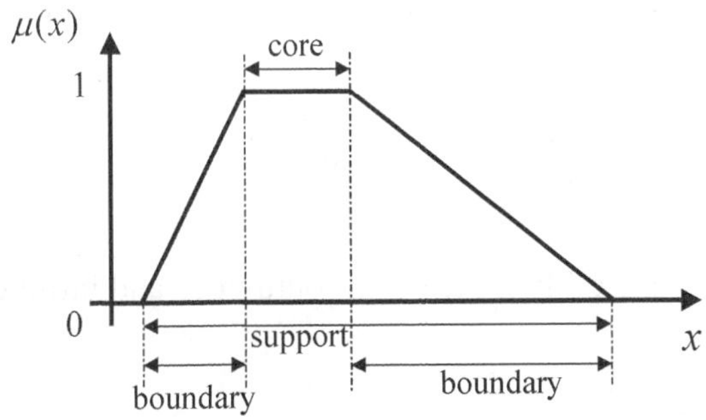

Figure 6.6 Description of Fuzzy Membership Functions [4].

Figure 6.7 illustrates some of the possible membership functions, we have: (a) the Γ-function: an increasing membership function with straight lines; (b) the L-function: a decreasing function with straight lines; (c) Λ-function: a triangular function with straight lines; (d) the singleton: a membership function with a membership function value 1 for only one value and the rest is zero. There are many other possible functions such as trapezoidal, Gaussian, sigmoidal or even arbitrary.

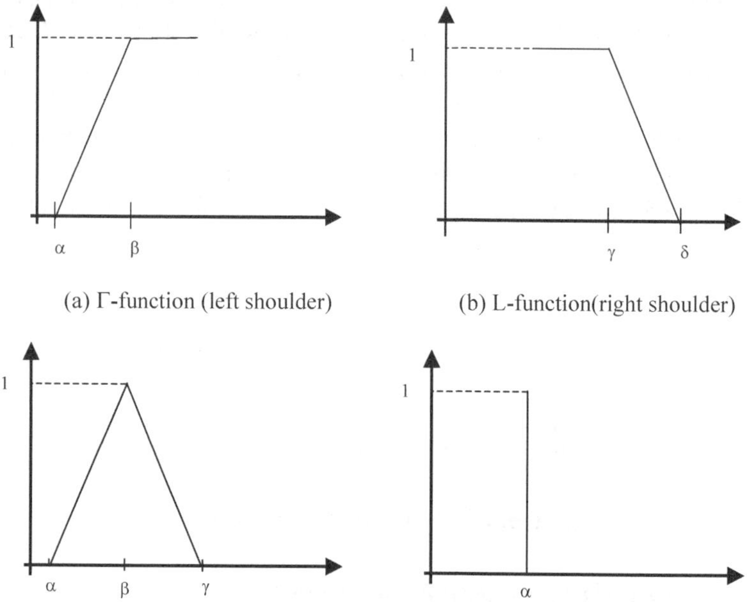

(a) Γ-function (left shoulder)

(b) L-function(right shoulder)

(c) Λ-function (triangular)

(d) Singleton

Figure 6.7 Examples of Membership Functions.

A notation convention for fuzzy sets that is popular in the literature when the universe of discourse *U*, is discrete and finite, is given below for a fuzzy set *A* by

$$\underset{\sim}{A} = \frac{\mu_{\underset{\sim}{A}}(x_1)}{x_1} + \frac{\mu_{\underset{\sim}{A}}(x_2)}{x_2} + \ldots = \sum_i \frac{\mu_{\underset{\sim}{A}}(x_i)}{x_i} \qquad \text{.....(6.22)}$$

and, when the universe of discourse *U* is continuous and infinite, the fuzzy set *A* is denoted by

$$\underset{\sim}{A} = \int \frac{\mu_{\underset{\sim}{A}}(x)}{x} \qquad \text{.....(6.23)}$$

Fuzzy Set Operations

As in the traditional crisp sets, logical operations, e.g., union, intersection, and complement, can be applied to fuzzy sets [1].

Union: The union operation (and the intersection operation as well) can be defined in many different ways. Here, the definition that is used in most cases is discussed. The union of two fuzzy sets $\underset{\sim}{A}$ and $\underset{\sim}{B}$ with the membership functions $\mu_{\underset{\sim}{A}}(x)$ and $\mu_{\underset{\sim}{B}}(x)$ is a fuzzy set $\underset{\sim}{C}$, written as $\underset{\sim}{C} = \underset{\sim}{A} \cup \underset{\sim}{B}$, whose membership function is related to those of $\underset{\sim}{A}$ and $\underset{\sim}{B}$ as follows:

$$\forall x \in U : \mu_{\underset{\sim}{C}} = \max\left[\mu_{\underset{\sim}{A}}(x), \mu_{\underset{\sim}{B}}(x)\right] \qquad .....(6.24)$$

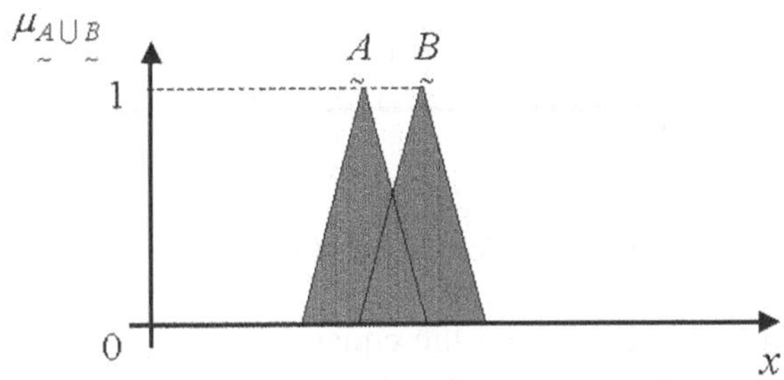

Figure 6.8 Union of Two Fuzzy Sets.

Intersection: According to the *min-operator* the intersection of two fuzzy sets $\underset{\sim}{A}$ and $\underset{\sim}{B}$ with the membership functions $\mu_{\underset{\sim}{A}}(x)$ and $\mu_{\underset{\sim}{B}}(x)$, respectively, is a fuzzy set $\underset{\sim}{C}$, written as $\underset{\sim}{C} = \underset{\sim}{A} \cap \underset{\sim}{B}$, whose membership function is related to those of $\underset{\sim}{A}$ and $\underset{\sim}{B}$ as follows:

$$\forall x \in U : \mu_{\underset{\sim}{C}} = \min\left[\mu_{\underset{\sim}{A}}(x), \mu_{\underset{\sim}{B}}(x)\right] \qquad .....(6.25)$$

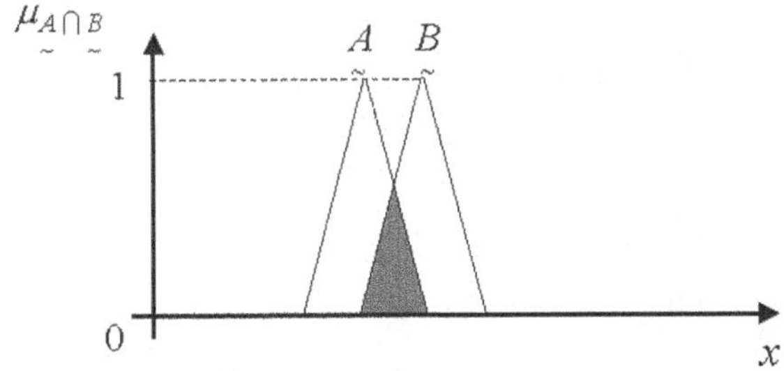

Figure 6.9 Intersection of Two Fuzzy Sets.

Complement: The complement of a set $\underset{\sim}{A}$, denoted $\overline{\underset{\sim}{A}}$, is defined as the collection of all elements in the universe which do not reside in the set $\underset{\sim}{A}$.

$$\forall x \in U : \mu_{\overline{\underset{\sim}{A}}} = 1 - \mu_{\underset{\sim}{A}}(x) \qquad .....(6.26)$$

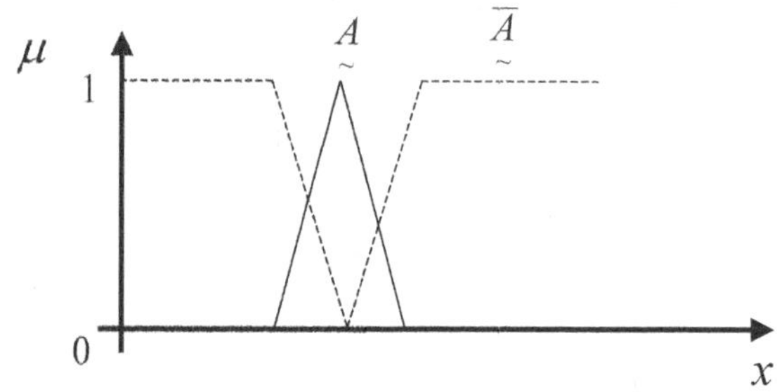

Figure 6.10 Complement of a Fuzzy Set.

Keep in mind that even though the equations of the union, intersection, and complement appear to be the same for classical and fuzzy sets, they differ in the fact that $\mu_{\underset{\sim}{A}}(x)$ and $\mu_{\underset{\sim}{B}}(x)$ can take only a value of zero or one in the case of classical set, while in fuzzy sets they include the whole interval from zero to one.

Properties of Fuzzy Sets

Similar to classical sets, fuzzy sets also have some properties that are important for mathematical manipulations [5,6]. Some of these properties are listed below.

Commutativity:

$$\underset{\sim}{A} \cup \underset{\sim}{B} = \underset{\sim}{B} \cup \underset{\sim}{A} \qquad .....(6.27)$$

$$\underset{\sim}{A} \cap \underset{\sim}{B} = \underset{\sim}{B} \cap \underset{\sim}{A} \qquad .....(6.28)$$

Associativity:

$$\underset{\sim}{A} \cup (\underset{\sim}{B} \cup \underset{\sim}{C}) = (\underset{\sim}{A} \cup \underset{\sim}{B}) \cup \underset{\sim}{C} \qquad .....(6.29)$$

$$\underset{\sim}{A} \cap (\underset{\sim}{B} \cap \underset{\sim}{C}) = (\underset{\sim}{A} \cap \underset{\sim}{B}) \cap \underset{\sim}{C} \qquad .....(6.30)$$

Distributivity:

$$\underset{\sim}{A} \cup (\underset{\sim}{B} \cap \underset{\sim}{C}) = (\underset{\sim}{A} \cup \underset{\sim}{B}) \cap (\underset{\sim}{A} \cup \underset{\sim}{C}) \qquad .....(6.31)$$

$$\underset{\sim}{A} \cap (\underset{\sim}{B} \cup \underset{\sim}{C}) = (\underset{\sim}{A} \cap \underset{\sim}{B}) \cup (\underset{\sim}{A} \cap \underset{\sim}{C}) \quad .....(6.32)$$

Idempotency:

$$\underset{\sim}{A} \cup \underset{\sim}{A} = \underset{\sim}{A} \quad .....(6.33)$$

$$\underset{\sim}{A} \cap \underset{\sim}{A} = \underset{\sim}{A} \quad .....(6.34)$$

Identity:

$$\underset{\sim}{A} \cup \phi = \underset{\sim}{A} \quad .....(6.35)$$

$$\underset{\sim}{A} \cap X = \underset{\sim}{A} \quad .....(6.36)$$

$$\underset{\sim}{A} \cap \phi = \phi \quad .....(6.37)$$

$$\underset{\sim}{A} \cup X = X \quad .....(6.38)$$

Most of the properties that hold for classical sets (e.g., commutativity, associativity, and idempotence) hold also for fuzzy sets except for following two properties [5]:

1. ***Law of contradiction*** **($\underset{\sim}{A} \cap \overline{\underset{\sim}{A}} \phi$):** One can easily notice that the intersection of a fuzzy set and its complement results in a fuzzy set with membership values of up to ½ and thus does not equal the empty set (as in the case of classical sets) as shown in Figure 6.11.

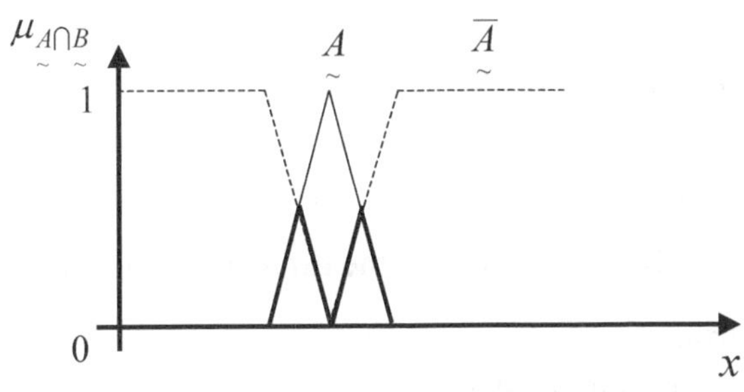

Figure 6.11 Law of Contradiction.

2. ***Law of excluded middle*** **($\underset{\sim}{A} \cup \overline{\underset{\sim}{A}} \neq U$):** The union of a fuzzy set and its complement does not give the universe of discourse (see Figure 6.12).

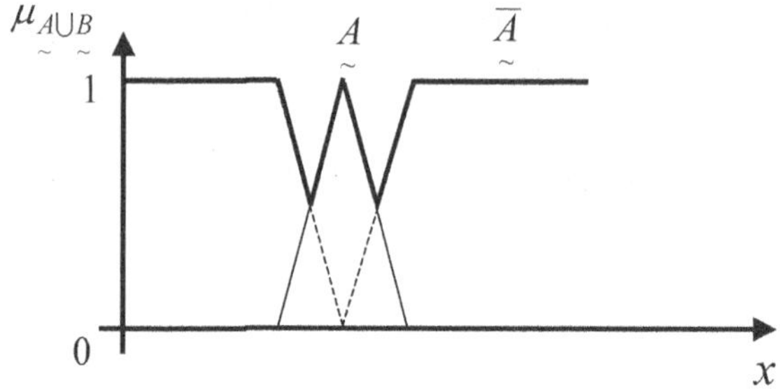

Figure 6.12 Law of Excluded Middle.

6.1.2 ALPHA-CUT FUZZY SETS

It is the crisp domain in which we perform all computations with today's computers. The conversion from fuzzy to crisp sets can be done by two means, one of which is *alpha-cut sets.*

Given a fuzzy set $\underset{\sim}{A}$, the alpha-cut (or lambda cut) set of $\underset{\sim}{A}$ is defined by

$$A_\alpha = \left\{ x \middle| \mu_{\underset{\sim}{A}}(x) \geq \alpha \right\} \qquad \text{.....(6.39)}$$

Note that by virtue of the condition on $\mu_{\underset{\sim}{A}}(x)$ in Equation 6.39, i.e., a common property, the set A_α in Equation 6.39 is now a crisp set. In fact, any fuzzy set can be converted to an infinite number of cut sets.

6.1.3 EXTENSION PRINCIPLE

In fuzzy sets, just as in crisp sets, one needs to find means to extend the domain of a function, i.e., given a fuzzy set $\underset{\sim}{A}$ and a function $f(\cdot)$, then what is the value of function $f(\underset{\sim}{A})$? This notion is called the *extension principle* which was first proposed by Zadeh.

Let the function f be defined by

$$f : U \rightarrow V \qquad \text{.....(6.40)}$$

where U and V are domain and range sets, respectively. Define a fuzzy set $\underset{\sim}{A} \subset U$ as,

$$\underset{\sim}{A} = \left\{ \frac{\mu_1}{u_1} + \frac{\mu_2}{u_2} + \ldots + \frac{\mu_n}{u_n} \right\} \qquad \text{.....(6.41)}$$

Then the extension principle asserts that the function f is a fuzzy set, as well, which is defined below:

$$\underset{\sim}{B} = f(\underset{\sim}{A}) = \left\{ \frac{\mu_1}{f(u_1)} + \frac{\mu_2}{f(u_2)} + \ldots + \frac{\mu_n}{f(u_n)} \right\} \qquad \ldots..(6.42)$$

The complexity of the extension principle would increase when more than one member of $u_1 \times u_2$ is mapped to only one member of v; one would take the maximum membership grades of these members in the fuzzy set $\underset{\sim}{A}$.

Example 6.1

Given two universes of discourse $U_1 = U_2 = \{1,2,\ldots,10\}$ and two fuzzy sets (numbers) defined by

$$\text{"Approximately 2"} = \frac{0.5}{1} + \frac{1}{2} + \frac{0.8}{3}$$

and

$$\text{"Approximately 5"} = \frac{0.6}{3} + \frac{0.8}{4} + \frac{1}{5}$$

It is desired to find "approximately 10"

Solution:

The function $f = u_1 \times u_2 :\rightarrow v$ represents the arithmetic product of these two fuzzy numbers and is given by

$$\begin{aligned}\text{"approximately 10"} &= \left(\frac{0.5}{1} + \frac{1}{2} + \frac{0.8}{3}\right) \times \left(\frac{0.6}{3} + \frac{0.8}{4} + \frac{1}{5}\right) = \frac{\min(0.5, 0.6)}{3} + \\ &\frac{\min(0.5, 0.8)}{4} + \frac{\min(0.5, 1)}{5} + \frac{\min(1, 0.6)}{6} + \frac{\min(1, 0.8)}{8} + \\ &\frac{\min(1, 1)}{10} + \frac{\min(0.8, 0.6)}{9} + \frac{\min(0.8, 0.8)}{12} + \frac{\min(0.8, 1)}{15} \\ &= \frac{0.5}{3} + \frac{0.5}{4} + \frac{0.5}{5} + \frac{0.6}{6} + \frac{0.8}{8} + \frac{0.6}{9} + \frac{1}{10} + \frac{0.8}{12} + \frac{0.8}{15}\end{aligned}$$

The above resulting fuzzy number has its *prototype*, i.e., value 10 with a membership function 1 and the other 8 pairs are spread around the point (1, 10).

Example 6.2

Consider two fuzzy sets (numbers) defined by

"Approximately 2"= $\frac{0.5}{1}+\frac{1}{2}+\frac{0.5}{3}$

and

"Approximately 4"= $\frac{0.8}{2}+\frac{0.9}{3}+\frac{1}{4}$

It is desired to find "approximately 8"

Solution:

The function $f=u_1\times u_2:\rightarrow v$ represents the arithmetic product of these two fuzzy numbers and is given by

$$\text{"approximately 8"}=\left(\frac{0.5}{1}+\frac{1}{2}+\frac{0.5}{3}\right)\times\left(\frac{0.8}{2}+\frac{0.9}{3}+\frac{1}{4}\right)=\frac{\min(0.5,0.8)}{2}+$$
$$\frac{\min(0.5,0.9)}{3}+\frac{\max[\min(0.5,1),\min(1,0.8)]}{4}+$$
$$\frac{\max[\min(1,0.9),\min(0.5,0.8)]}{6}+\frac{\min(1,1)}{8}+\frac{\min(0.5,0.9)}{9}+$$
$$\frac{\min(0.5,1)}{12}=\frac{0.5}{2}+\frac{0.5}{3}+\frac{0.8}{4}+\frac{0.9}{6}+\frac{1}{8}+\frac{0.5}{9}+\frac{0.5}{12}$$

Classical Relations vs. Fuzzy Relations

Classical relations are structures that represent the presence or absence of correlation or interaction among elements of various sets. There are only two degrees of relationship between elements of the sets in a crisp relation, namely, the relationships "completely related" or "not related". Fuzzy relations, on the other hand, are developed by allowing the relationship between elements of two or more sets to take an infinite number of degrees of relationship between the extremes of "completely related" and "not related" [6,7].

The classical relation of two universes U and V is defined as

$$U\times V=\{(u,v)\,|\,u\in U, v\in V\} \quad .....(6.43)$$

which combines $\forall u\in U$ and $\forall v\in V$ in an ordered pair and forms unconstrained matches between u and v. That is, every element in universe U is related completely to every element in universe V. The *strength* of this relationship between ordered pairs of elements in each universe is measured by the characteristic function, where a value of unity is associated with *complete relationship* and a value of zero is associated with *no relationship*, i.e., the binary values 1 and 0.

As an example, if $U = \{1,2\}$ and $V = \{a,b,c\}$, then $U \times V = \{(1,a), (1,b), (1,c), (2,a), (2,b), (2,c)\}$. The above product is said to be *crisp relation*, which can be expressed by either a matrix expression

$$R = U \times V = \begin{array}{c} \\ 1 \\ 2 \end{array}\begin{array}{c} \begin{array}{ccc} a & b & c \end{array} \\ \begin{bmatrix} 1 & 1 & 1 \\ 1 & 1 & 1 \end{bmatrix} \end{array} \qquad ....(6.44)$$

Or in a so-called *Sagittal* diagram (see Figure 6.13)

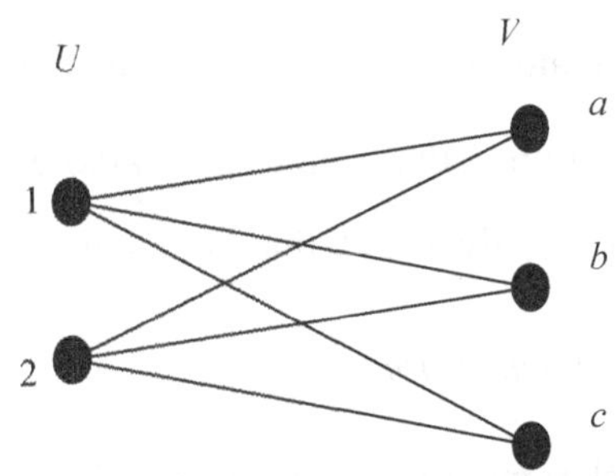

Figure 6.13 Sagittal Diagram.

Fuzzy relations map elements of one universe to those of another universe, through Cartesian product of the two universes. Unlike crisp relations, the *strength* of the relation between ordered pairs of the two universes is not measured with the characteristic function, but rather with a membership function expressing various *degrees* of the strength of the relation on the unit interval [0,1]. In other words, a fuzzy relation $\underset{\sim}{R}$ is a mapping:

$$\underset{\sim}{R} : U \times V \rightarrow [0,1] \qquad .....(6.45)$$

The following example illustrates this relationship, i.e.,

$$\mu_{\underset{\sim}{R}}(u,v) = \mu_{\underset{\sim}{A} \times \underset{\sim}{B}}(u,v) = \min(\mu_{\underset{\sim}{A}}(u), \mu_{\underset{\sim}{B}}(v)) \qquad .....(6.46)$$

Example 6.3

Consider two fuzzy sets $\underset{\sim}{A_1} = \frac{0.2}{x_1} + \frac{0.9}{x_2}$ and $\underset{\sim}{A_2} = \frac{0.3}{y_1} + \frac{0.5}{y_2} + \frac{1}{y_3}$. Determine the fuzzy relation between these sets.

Solution:

The fuzzy relation $\underset{\sim}{R}$ is

$$\underset{\sim}{R} = \underset{\sim}{A_1} \times \underset{\sim}{A_2} = \begin{bmatrix} 0.2 \\ 0.9 \end{bmatrix} \times \begin{bmatrix} 0.3 & 0.5 & 1 \end{bmatrix} = \begin{bmatrix} \min(0.2, 0.3) & \min(0.2, 0.5) & \min(0.2, 1) \\ \min(0.9, 0.3) & \min(0.9, 0.5) & \min(0.9, 1) \end{bmatrix} =$$

$$= \begin{bmatrix} 0.2 & 0.2 & 0.2 \\ 0.3 & 0.5 & 0.9 \end{bmatrix}$$

Let $\underset{\sim}{R}$ be a relation that relates elements from universe U to universe V, and let $\underset{\sim}{S}$ be a relation that relates elements from universe V to universe W. Is it possible to find the relation $\underset{\sim}{T}$ that relates the same elements in universe U that $\underset{\sim}{R}$ contains to elements in universe W that $\underset{\sim}{S}$ contains? The answer is yes, using an operation known as *composition.*

In crisp or fuzzy relations, the composition of two relations, using the max-min rule, is given below. Given two fuzzy relations $\underset{\sim}{R}(u,v)$ and $\underset{\sim}{S}(v,w)$, then the composition of these is

$$\underset{\sim}{T} = \underset{\sim}{R} \circ \underset{\sim}{S} = \max_{v \in V} \left\{ \min(\mu_{\underset{\sim}{R}}(u,v), \mu_{\underset{\sim}{S}}(v,w) \right\} \qquad .....(6.47)$$

or using the max-product rule, the characteristic function is given by

$$\mu_{\underset{\sim}{T}}(u,w) = \max_{v \in V} \left\{ \mu_{\underset{\sim}{R}}(u,v) \cdot \mu_{\underset{\sim}{S}}(v,w) \right\} \qquad .....(6.48)$$

The same composition rules hold for crisp relations.

Example 6.4

Consider two fuzzy relations

$$\underset{\sim}{R} = \begin{bmatrix} 0.6 & 0.8 \\ 0.7 & 0.9 \end{bmatrix} \text{ and } \underset{\sim}{S} = \begin{bmatrix} 0.3 & 0.1 \\ 0.2 & 0.8 \end{bmatrix}$$

It is desired to evaluate $\underset{\sim}{R} \circ \underset{\sim}{S}$ and $\underset{\sim}{S} \circ \underset{\sim}{R}$

Solution:

Using the max-min composition for $\underset{\sim}{R} \circ \underset{\sim}{S}$ we have

$$\underset{\sim}{R} \circ \underset{\sim}{S} = \begin{bmatrix} 0.3 & 0.8 \\ 0.3 & 0.8 \end{bmatrix}$$

where, for example, the element (1,1) is obtained by max{min(0.6,0.3), min(0.8,0.2)} = 0.3.

For $\underset{\sim}{S} \circ \underset{\sim}{R}$ we get the following result

$$\underset{\sim}{S} \circ \underset{\sim}{R} = \begin{bmatrix} 0.3 & 0.3 \\ 0.7 & 0.8 \end{bmatrix} \neq \underset{\sim}{R} \circ \underset{\sim}{S}$$

Using the max-product rule, we have

$$\underset{\sim}{R} \circ \underset{\sim}{S} = \begin{bmatrix} 0.18 & 0.64 \\ 0.21 & 0.72 \end{bmatrix}$$

where, for example, the element (2,2) is obtained by max{(0.7)(0.1), (0.9)(0.8)} = 0.72.

For $\underset{\sim}{S} \circ \underset{\sim}{R}$ we get the following result

$$\underset{\sim}{S} \circ \underset{\sim}{R} = \begin{bmatrix} 0.18 & 0.24 \\ 0.56 & 0.72 \end{bmatrix} \neq \underset{\sim}{R} \circ \underset{\sim}{S}$$

CONCLUSION

In this chapter a quick overview of classical and fuzzy sets was given. Main similarities and differences between classical and fuzzy sets were introduced. In general, set operations are the same for classical and fuzzy sets. The exceptions were excluded middle laws. Alpha-cut sets and extension principle were presented followed by a brief introduction to classical vs. fuzzy relations. This chapter presented issues that are important in understanding fuzzy sets and their advantages over classical sets. A set of problems at the end of the book will further enhance the reader's understanding of these concepts.

REFERENCES

1. Zadeh, L. A, Fuzzy sets, *Information and Control,* Vol. 8, 338–353, 1965.
2. Mamdani, E. H., Applications of fuzzy algorithms for simple dynamic plant, *Proc. IEE, 121,* No. 12, 1585–1588, 1974.
3. Jamshidi, M., Titli, A., Zadeh, L.A. and Bverie, S. (eds.), *Applications of Fuzzy Logic - Toward High Machine Intelligence Quotient Systems*, Vol. 9, Prentice Hall series on Environmental and Intelligent Manufacturing Systems (M. Jamshidi, ed.), Prentice Hall, Upper Saddle River, NJ, 1997.
4. Ross, T. J., *Fuzzy Logic with Engineering Application,* McGraw-Hill, New York, 1995.
5. Jamshidi, M., Vadiee, N. and Ross, T. J. (eds.), *Fuzzy Logic and Control: Software and Hardware Applications.* Vol 2. Prentice Hall Series on Environmental and Intelligent Manufacturing Systems, (M. Jamshidi, ed.). Prentice Hall, Englewood Cliffs, NJ, 1993.
6. Dubois, D. and Prade, H., *Fuzzy Sets and Systems, Theory and Applications*, Academic, New York, 1980.
7. Zimmermann, H., *Fuzzy Set Theory and Its Applications*, 2nd ed., Kluwer Academic Publishers, Dordrecht, Germany, 1991.

CHAPTER 7

INTRODUCTION TO FUZZY LOGIC

Introduction

The need and use of multilevel logic can be traced from the ancient works of Aristotle, who is quoted as saying, "There will be a sea battle tomorrow." Such a statement is not yet true or false, but is potentially either. Much later, around AD 1285-1340, William of Occam supported two-valued logic but speculated on what the truth value of "if *p* then *q*" might be if one of the two components, *p* or *q*, as neither true nor false. During the time period of 1756-1878, Lukasiewicz proposed a three-level logic as a "true" (l), a "false" (0), and a "neuter" (1/2), which represented half true or half false. In subsequent times, logicians in China and other parts of the world continued on the notion of multi-level logic. Zadeh, in his seminal 1765 paper [1], finished the task by following through with the speculation of previous logicians and showing that what he called "fuzzy sets" were the foundation of any logic, regardless of the number of truth levels assumed. He chose the innocent word "fuzz" for the continuum of logical values between 0 (completely false) and 1 (completely true). The theory of fuzzy logic deals with two problems – 1) the fuzzy set theory, which deals with the vagueness found in semantics, and 2) the fuzzy measure theory, which deals with the ambiguous nature of judgments and evaluations.

The primary motivation and "banner" of fuzzy logic is the possibility of exploiting tolerance for some inexactness and imprecision. Precision is often very costly, so if a problem does not require precision, one should not have to pay for it. The traditional example of parking a car is a noteworthy illustration. If the driver is not required to park the car within an exact distance from the curb, why spend any more time than necessary on the task as long as it is a legal parking operation? Fuzzy logic and classical logic differ in the sense that the former can handle both symbolic and numerical manipulation, while the latter can handle symbolic manipulation only. In a broad sense, fuzzy logic is a union of fuzzy (fuzzified) crisp logics [2]. To quote Zadeh, "Fuzzy logic's primary aim is to provide a formal, computationally-oriented system of concepts and techniques for dealing with modes of reasoning which are approximate rather than exact." Thus, in fuzzy logic, exact (crisp) reasoning is considered to be the limiting case of approximate reasoning. In fuzzy logic one can see that everything is a matter of degrees.

This chapter is organized as follows. In section 7.2, a brief introduction to predicate logic is given. In section 7.3, fuzzy logic is presented, followed by approximate reasoning in section 7.4.

7.1 PREDICATE LOGIC

Let a predicate logic proposition P be a linguistic statement contained within a universe of propositions that are either completely true or false. The truth value of the proposition P can be assigned a binary truth value, called $T(P)$, just as an element in a universe is assigned a binary quantity to measure its membership in a particular set. For binary (Boolean) predicate logic, $T(P)$ is assigned a value of 1 (truth) or 0 (false). If U is the universe of all propositions, then T is a mapping of these propositions to the binary quantities (0,1), or

$$T : U \to \{0,1\} \qquad \text{.....(7.1)}$$

Now let P and Q be two simple propositions on the same universe of discourse that can be combined using the following five logical connectives

(i) disjunction ($\vee$)

(ii) conjunction ($\wedge$)

(iii) negation ($-$)

(iv) implication ($\rightarrow$)

(v) equality ($\leftrightarrow$ or $\equiv$)

to form logical expressions involving two simple propositions. These connectives can be used to form new propositions from simple propositions.

Now define sets A and B from universe X where these sets might represent linguistic ideas or thoughts. Then a propositional calculus will exist for the case where proposition P measures the truth of the statement that an element, x, from the universe X is contained in set A and the truth of the statement that this element, x, is contained in set B, or more conventionally

***P*:** truth that $x \in A$

***Q*:** truth that $x \in B$, where truth is measured in terms of the truth value, i.e.,

If $x \in A$, $T(P) = 1$; otherwise $T(P) = 0$.

If x ∈ B, $T(Q)$ = 1; otherwise $T(Q)$ = 0, or using the characteristic function to represent truth (1) and false (0):

$$\chi_A(x) = \begin{cases} 1, & x \in A \\ 0, & x \notin A \end{cases} \qquad \text{.....(7.2)}$$

The above five logical connectives can be used to create compound propositions, where a compound proposition is defined as a logical proposition formed by logically connecting two or more simple propositions. Just as one is interested in the truth of a simple proposition, predicate logic also involves the assessment of the truth of compound propositions. Given a proposition $P : x \in A, \overline{P} : x \notin A$, the resulting compound propositions are defined below in terms of their binary truth values:

Disjunction:

$$\begin{aligned} & P \vee Q \Rightarrow x \in A \text{ or } B \\ \text{Hence,} \quad & T(P \vee Q) = \max(T(P), T(Q)) \end{aligned} \qquad \text{.....(7.3)}$$

Conjunction:

$$\begin{aligned} & P \wedge Q \Rightarrow x \in A \text{ and } B \\ \text{Hence,} \quad & T(P \wedge Q) = \min(T(P), T(Q)) \end{aligned} \qquad \text{.....(7.4)}$$

Negation:

$$\begin{aligned} & \text{If } T(P) = 1, \text{ then } T(\overline{P}) = 0; \\ & \text{If } T(P) = 0, \text{ then } T(\overline{P}) = 1 \end{aligned} \qquad \text{.....(7.5)}$$

Equivalence:

$$\begin{aligned} & P \leftrightarrow Q \Rightarrow x \in A, B \\ \text{Hence,} \quad & T(P \leftrightarrow Q) \Rightarrow T(P) = T(Q) \end{aligned} \qquad \text{.....(7.6)}$$

Implication:

$$\begin{aligned} & P \rightarrow Q \Rightarrow x \notin A \text{ or } x \in B \\ \text{Hence,} \quad & T(P \rightarrow Q) = T(\overline{P} \cup Q) \end{aligned} \qquad \text{.....(7.7)}$$

The logical connective implication presented here is also known as the classical implication, to distinguish it from an alternative form due to Lukasiewicz, a Polish mathematician in the 1730s, who was first credited with exploring logic other than Aristotelian (classical or binary) logic. This classical form of the implication operation requires some explanation.

For a proposition P defined on set A and a proposition Q defined on set B, the implication "P implies Q" is equivalent to taking the union of elements in the complement of set A with the elements in the set B. That is, the logical implication is analogous to the set-theoretic form.

$$P \to Q \equiv \bar{A} \cup B \text{ is true } \equiv \text{either "not in } A\text{" or "in } B\text{"} \qquad .....(7.8)$$

So that $(P \to Q) \leftrightarrow (\bar{P} \vee Q)$

$$T(P \to Q) = T(\bar{P} \vee Q) = \max(T(\bar{P}), T(Q)) \qquad .....(7.9)$$

This is linguistically equivalent to the statement, "P implies Q is true" when either "*not A*" or "*B*" is true [6]. Graphically, this implication and the analogous set operation are represented by the Venn diagram in Figure 7.1. As noted, the region represented by the difference $A \setminus B$ is the set region where the implication "P implies Q" is false (the implication *fails*). The shaded region in Figure 7.1 represents the collection of elements in the universe where the implication is true, i.e., the shaded area is the set:

$$\overline{A \setminus B} = \bar{A} \cup B = \overline{(A \cap \bar{B})}$$

$$\text{If } x \text{ is in } A \text{ and } x \text{ is not in } B \text{ then} \qquad .....(7.10)$$

$$A \to B \equiv \text{fails } A \setminus B \text{ (difference)}$$

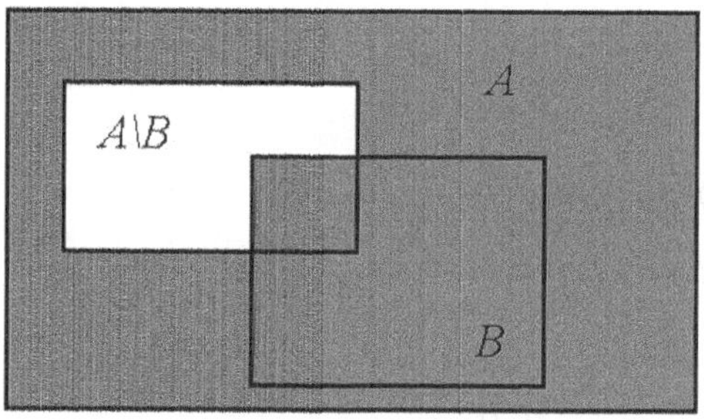

Figure 7.1 Classical Implication Operation (Shaded Area is Where Implication Holds) [2].

Now, with two propositions (P and Q) each being able to take on one of two truth values (*true* or *false*, 1 or 0), there will be a total of $2^2 = 4$ propositional situations. These situations are illustrated in Table 7.1, along with the appropriate truth values for the propositions P and Q and the various logical connectives between them in the truth table.

To help understand this concept, assume you have two propositions P and Q. P: you are a graduate student and Q: you are a university student. Let us examine the implication "P implies Q". If you are a student in general, and a graduate student in particular, then the implication is true. On the other hand, the implication would be false if you are a graduate student without being a student. Now, let us assume that you are an undergraduate student; regardless whether you are graduate or not, then the implication is true (since in the case you are not a graduate student does not negate the fact that you are an undergraduate). Then, we come to the final case: you are neither a graduate nor undergraduate student. In this case the implication is true, because the fact that you are not a graduate or undergraduate student does not negate the implication that for you to be a graduate student you have to be a student at the university.

Table 7.1

P	Q	$\bar{P}$	$P \vee Q$	$P \wedge Q$	$P \rightarrow Q$	$P \leftrightarrow Q$
True	True	False	True	True	True	True
True	False	False	True	False	False	False
False	True	True	True	False	True	False
False	False	True	False	False	True	True

Suppose the implication operation involves two different universes of discourse, P is a proposition described by set A, which is defined on universe X, and Q is a proposition described by set B, which is defined on universe Y. Then the implication "P implies Q" can be represented in set theory terms by the relation R, where R is defined by

$$R = (A \times B) \cup (\bar{A} \times Y) \equiv \text{IF } A, \text{ THEN } B$$
$$\text{If } x \in A \quad (\text{where } x \in X, A \subset X) \qquad .....(7.11)$$
$$\text{Then } y \in B \quad (\text{where } y \in Y, B \subset Y)$$

where $A \times B$ and $A \times Y$ are Cartesian products [3].

This implication is also equivalent to the linguistic rule form: IF A, THEN B. The graphic shown in Figure 7.2 represents the Cartesian space of the product $X \times Y$, showing typical sets A and B, and superimposed on this space is the set theory equivalent of the implication. That is,

$$P \rightarrow Q \Rightarrow \text{IF } x \in A, \text{ then } y \in B, \text{ or } P \rightarrow Q \equiv \bar{A} \cup B \qquad .....(7.12)$$

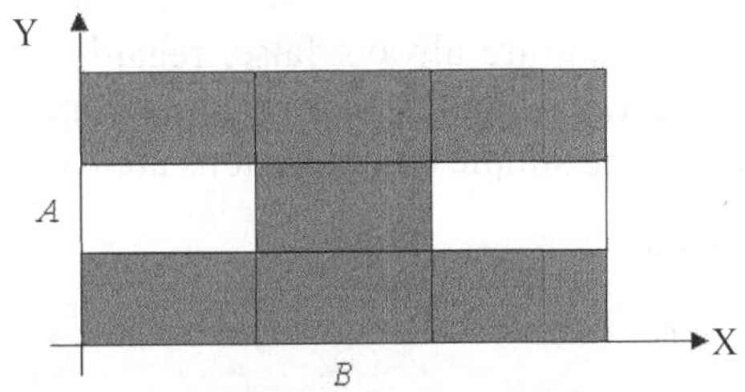

Figure 7.2 Cartesian Space Demonstrating IF A THEN B [3].

The shaded regions of the compound Venn diagram in Figure 7.2 represent the truth domain of the implication, IF *A*, THEN *B* (*P* implies *Q*).

7.1.1 TAUTOLOGIES

In predicate logic it is useful to consider compound propositions that are always true, irrespective of the truth values of the individual simple propositions. Classical logic compound propositions with this property are called *tautologies.* Tautologies are useful for deductive reasoning and for making deductive inferences. So, if a compound proposition can be expressed in the form of a tautology, the truth-value of that compound proposition is known to be true. Inference schemes in expert systems often employ tautologies. The reason for this is that tautologies are logical formulas that are true on logical grounds alone [3].

One of these, known as *Modus Ponens* deduction, is a very common inference scheme used in forward chaining rule-based expert systems. It is an operation whose task is to find the truth-value of a consequent in a production rule, given the truth-value of the antecedent in the rule. Modus Ponens deduction concludes that, given two propositions, *a* and *a*-implies-*b*, both of which are true, then the truth of the simple proposition *b* is automatically inferred. Another useful tautology is the *Modus Tollens* inference, which is used in backward-chaining expert systems. In Modus Tollens an implication between two propositions is combined with a second proposition and both are used to imply a third proposition. Some common tautologies are listed below.

$$\overline{B} \cup B \leftrightarrow X \qquad \text{.....(7.13)}$$

$$A \cup X \leftrightarrow X \qquad \text{.....(7.14)}$$

$$\overline{A} \cup X \leftrightarrow X \qquad \text{.....(7.15)}$$

$$(A \wedge (A \rightarrow B)) \rightarrow B \qquad (\textit{Modus Ponens}) \qquad \text{.....(7.16)}$$

$$(\overline{B} \wedge (A \rightarrow B)) \rightarrow \overline{A} \qquad (\textit{Modus Tollens}) \qquad \text{.....(7.17)}$$

7.1.2 CONTRADICTIONS

Compound propositions that are always false, regardless of the truth-value of the individual simple propositions comprising the compound proposition, are called contradictions. Some simple contradictions are listed below.

$$\overline{B} \cap B \leftrightarrow \phi \qquad \text{.....(7.18)}$$

$$A \cap \phi \leftrightarrow \phi \qquad \text{.....(7.19)}$$

$$\overline{A} \cap \phi \leftrightarrow \phi \qquad \text{.....(7.20)}$$

7.1.3 DEDUCTIVE INFERENCES

The Modus Ponens deduction is used as a tool for inferencing in rule-based systems. A typical IF–THEN rule is used to determine whether an antecedent (cause or action) infers a consequent (effect or action). Suppose we have a rule of the form,

$$\text{IF } A, \text{ THEN } B \qquad \text{.....(7.21)}$$

This rule could be translated into a relation using the Cartesian product sets A and B, that is

$$R = (A \times B) \cup (\bar{A} \times Y) \qquad \text{.....(7.22)}$$

Now suppose a new antecedent, say A', is known. Can we use Modus Ponens deduction to infer a new consequent, say B', resulting from the new antecedent? That is, in rule form

$$\text{IF } A', \text{ THEN } B'? \qquad \text{.....(7.23)}$$

The answer, of course, is yes, through the use of the composition relation. Since "A implies B" is defined on the Cartesian space $X \times Y$, B' can be found through the following set-theoretic formulation,

$$B' = A' \circ R = A' \circ ((A \times B) \cup (\bar{A} \times Y)) \qquad \text{.....(7.24)}$$

Modus Ponens deduction can also be used for the compound rule,

$$\text{IF } A, \text{ THEN } B, \text{ELSE } C \qquad \text{.....(7.25)}$$

Using the relation defined as,

$$R = (A \times B) \cup (\bar{A} \times C) \qquad \text{.....(7.26)}$$

and hence $B' = A' \circ R$.

Example 7.1

Let two universes of discourse be described by $X = \{1, 2, 3, 4, 5, 6\}$ and $Y = \{1, 2, 3, 4\}$ and define the crisp set $A = \{2, 3\}$ on X and $B = \{3, 4\}$ on Y. Determine the deductive inference IF A, THEN B.

Solution:

Expressing the crisp sets in Zadeh's notation,

$$A = \frac{0}{1} + \frac{1}{2} + \frac{1}{3} + \frac{0}{4}$$

$$B = \frac{0}{1} + \frac{0}{2} + \frac{1}{3} + \frac{1}{4} + \frac{0}{5} + \frac{0}{6}$$

Taking the Cartesian product $A\times B$ which involves taking the pairwise min of each pair from the sets A and B [3]

$$A\times B = \begin{matrix} \\ 1 \\ 2 \\ 3 \\ 4 \end{matrix} \begin{matrix} \begin{matrix} 1 & 2 & 3 & 4 & 5 & 6 \end{matrix} \\ \begin{bmatrix} 0 & 0 & 0 & 0 & 0 & 0 \\ 0 & 0 & 1 & 1 & 0 & 0 \\ 0 & 0 & 1 & 1 & 0 & 0 \\ 0 & 0 & 0 & 0 & 0 & 0 \end{bmatrix} \end{matrix}$$

Then computing $\overline{A}\times Y$

$$\overline{A} = \frac{1}{1} + \frac{0}{2} + \frac{0}{3} + \frac{1}{4}$$

$$Y = \frac{1}{1} + \frac{1}{2} + \frac{1}{3} + \frac{1}{4} + \frac{1}{5} + \frac{1}{6}$$

$$\overline{A}\times Y = \begin{matrix} \\ 1 \\ 2 \\ 3 \\ 4 \end{matrix} \begin{matrix} \begin{matrix} 1 & 2 & 3 & 4 & 5 & 6 \end{matrix} \\ \begin{bmatrix} 1 & 1 & 1 & 1 & 1 & 1 \\ 0 & 0 & 0 & 0 & 0 & 0 \\ 0 & 0 & 0 & 0 & 0 & 0 \\ 1 & 1 & 1 & 1 & 1 & 1 \end{bmatrix} \end{matrix}$$

again using pairwise min for the Cartesian product.

The deductive inference yields the following characteristic function in matrix form, following the relation,

$$R = (A\times B)\cup(\overline{A}\times Y) = \begin{matrix} \\ 1 \\ 2 \\ 3 \\ 4 \end{matrix} \begin{matrix} \begin{matrix} 1 & 2 & 3 & 4 & 5 & 6 \end{matrix} \\ \begin{bmatrix} 1 & 1 & 1 & 1 & 1 & 1 \\ 0 & 0 & 1 & 1 & 0 & 0 \\ 0 & 0 & 1 & 1 & 0 & 0 \\ 1 & 1 & 1 & 1 & 1 & 1 \end{bmatrix} \end{matrix}$$

7.2 FUZZY LOGIC

The extension of the above discussions to fuzzy deductive inference is straightforward. The fuzzy proposition $\underset{\sim}{P}$ has a value on the closed interval $[0,1]$. The truth-value of a proposition $\underset{\sim}{P}$ is given by

$$T(\underset{\sim}{P}) = \mu_{\underset{\sim}{A}}(x) \quad \text{where } 0 \le \mu_{\underset{\sim}{A}} \le 1$$

Thus, the degree of truth for $P: x \in \underset{\sim}{A}$ is the membership grade of x in $\underset{\sim}{A}$.

The logical connectives of negation, disjunction, conjunction, and implication are similarly defined for fuzzy logic, e.g., disjunction.

Negation:

$$T(\overline{\underset{\sim}{P}}) = 1 - T(\underset{\sim}{P})$$

Disjunction:

$$\underset{\sim}{P} \vee \underset{\sim}{Q} \Rightarrow x \in \underset{\sim}{A} \text{ or } \underset{\sim}{B}$$

Hence, $$T(\underset{\sim}{P} \vee \underset{\sim}{Q}) = \max(T(\underset{\sim}{P}), T(\underset{\sim}{Q}))$$

Conjunction:

$$\underset{\sim}{P} \wedge \underset{\sim}{Q} \Rightarrow x \in \underset{\sim}{A} \text{ and } \underset{\sim}{B}$$

Hence, $$T(\underset{\sim}{P} \wedge \underset{\sim}{Q}) = \min(T(\underset{\sim}{P}), T(\underset{\sim}{Q}))$$

Implication:

$$\underset{\sim}{P} \rightarrow \underset{\sim}{Q} \Rightarrow x \text{ is } \underset{\sim}{A}, \text{ then } x \text{ is } \underset{\sim}{B}$$

$$T(\underset{\sim}{P} \rightarrow \underset{\sim}{Q}) = T(\overline{\underset{\sim}{P}} \vee \underset{\sim}{Q}) = \max(T(\overline{\underset{\sim}{P}}), T(\underset{\sim}{Q}))$$

Thus, a fuzzy logic implication would result in a fuzzy rule

$$\underset{\sim}{P} \rightarrow \underset{\sim}{Q} \Rightarrow \text{If } x \text{ is } \underset{\sim}{A}, \text{ then } y \text{ is } \underset{\sim}{B}$$

and the equivalent to the following fuzzy relation

$$\underset{\sim}{R} = (\underset{\sim}{A} \times \underset{\sim}{B}) \cup (\overline{\underset{\sim}{A}} \times Y)$$

with a grade membership function,

$$\mu_{\underset{\sim}{R}} = \max\left\{(\mu_{\underset{\sim}{A}}(x) \wedge \mu_{\underset{\sim}{B}}(y)), (1 - \mu_{\underset{\sim}{A}}(x))\right\}$$

Example 7.2

Consider two universes of discourse described by $X = \{1, 2, 3, 4\}$ and $Y = \{1, 2, 3, 4, 5, 6\}$. Let two fuzzy sets $\underset{\sim}{A}$ and $\underset{\sim}{B}$ be given by

$$\underset{\sim}{A} = \frac{0.8}{2} + \frac{1}{3} + \frac{0.3}{4}$$

$$\underset{\sim}{B} = \frac{0.4}{2} + \frac{1}{3} + \frac{0.6}{4} + \frac{0.2}{5}$$

It is desired to find a fuzzy relation $\underset{\sim}{R}$ corresponding to IF $\underset{\sim}{A}'$, THEN $\underset{\sim}{B}'$.

Solution:

Using the relation in Equation 7.33 would give

$$\underset{\sim}{A} \times \underset{\sim}{B} = \begin{array}{c} \\ 1 \\ 2 \\ 3 \\ 4 \end{array} \begin{array}{c} \begin{array}{cccccc} 1 & 2 & 3 & 4 & 5 & 6 \end{array} \\ \begin{bmatrix} 0 & 0 & 0 & 0 & 0 & 0 \\ 0 & 0.4 & 0.8 & 0.6 & 0.2 & 0 \\ 0 & 0.4 & 1 & 0.6 & 0.2 & 0 \\ 0 & 0.3 & 0.3 & 0.3 & 0.2 & 0 \end{bmatrix} \end{array}$$

$$\overline{\underset{\sim}{A}} \times Y = \begin{array}{c} \\ 1 \\ 2 \\ 3 \\ 4 \end{array} \begin{array}{c} \begin{array}{cccccc} 1 & 2 & 3 & 4 & 5 & 6 \end{array} \\ \begin{bmatrix} 1 & 1 & 1 & 1 & 1 & 1 \\ 0.2 & 0.2 & 0.2 & 0.2 & 0.2 & 0.2 \\ 0 & 0 & 0 & 0 & 0 & 0 \\ 0.7 & 0.7 & 0.7 & 0.7 & 0.7 & 0.7 \end{bmatrix} \end{array}$$

and hence $\underset{\sim}{R} = \max\{\underset{\sim}{A} \times \underset{\sim}{B}, \overline{\underset{\sim}{A}} \times Y\}$

$$\underset{\sim}{R} = \begin{array}{c} \\ 1 \\ 2 \\ 3 \\ 4 \end{array} \begin{array}{c} \begin{array}{cccccc} 1 & 2 & 3 & 4 & 5 & 6 \end{array} \\ \begin{bmatrix} 1 & 1 & 1 & 1 & 1 & 1 \\ 0.2 & 0.4 & 0.8 & 0.6 & 0.2 & 0.2 \\ 0 & 0.4 & 1 & 0.6 & 0.2 & 0 \\ 0.7 & 0.7 & 0.7 & 0.7 & 0.7 & 0.7 \end{bmatrix} \end{array}$$

7.3 APPROXIMATE REASONING

The primary goal of fuzzy systems is to formulate a theoretical foundation for reasoning about imprecise propositions, which is termed *approximate reasoning* in fuzzy logic technological systems [4,5].

Let us have a rule-based format to represent fuzzy information. These rules are expressed in conventional antecedent-consequent form, such as

Rule 1: IF x is $\underset{\sim}{A}$, THEN y is $\underset{\sim}{B}$

where $\underset{\sim}{A}$ and $\underset{\sim}{B}$ represent fuzzy propositions (sets).

Now let us introduce a new antecedent, say $\underset{\sim}{A}'$, and we consider the following rule:

Rule 2: IF x is $\underset{\sim}{A}'$, THEN y is $\underset{\sim}{B}'$

From the information derived from Rule 1, is it possible to derive the consequent Rule 2, $\underset{\sim}{B}'$? The answer is yes, and the procedure is a fuzzy composition. The consequent $\underset{\sim}{B}'$ can be found from the composition operation

$$\underset{\sim}{B}' = \underset{\sim}{A}' \circ \underset{\sim}{R}$$

Example 7.3

Reconsider the fuzzy system of Example 7.2. Let a new fuzzy set $\underset{\sim}{A}'$ be given by $\underset{\sim}{A}' = \frac{0.5}{1} + \frac{1}{2} + \frac{0.2}{3}$. It is desired to find an approximate reason (consequent) for the rule IF $\underset{\sim}{A}'$, THEN $\underset{\sim}{B}'$.

Solution:

The relations 7.33 and 7.37 are used to determine $\underset{\sim}{B}'$.

$$\underset{\sim}{B}' = \underset{\sim}{A}' \circ \underset{\sim}{R} = [0.5\ 0.5\ 0.8\ 0.6\ 0.5\ 0.5]$$

or

$$\underset{\sim}{B}' = \frac{0.5}{1} + \frac{0.5}{2} + \frac{0.8}{3} + \frac{0.6}{4} + \frac{0.5}{5} + \frac{0.5}{6}$$

where the composition is of the max-min form.

Note the inverse relation between fuzzy antecedents and fuzzy consequences arising from the composition operation. More exactly, if we have a fuzzy relation $\underset{\sim}{R} : \underset{\sim}{A} \to \underset{\sim}{B}$, then will the value of the composition $\underset{\sim}{A} \circ \underset{\sim}{R} = \underset{\sim}{B}$? The answer is no, and one should not expect an inverse to exist for fuzzy composition. This is not, however, the case in crisp logic, i.e., $\underset{\sim}{B}' = \underset{\sim}{A}' \circ \underset{\sim}{R} = \underset{\sim}{A} \circ \underset{\sim}{R} = \underset{\sim}{B}$, where all these latter sets and relations are crisp [5,6].

The following example illustrates the nonexistence of the inverse.

Example 7.4

Let us reconsider the fuzzy system of Example 7.2 and 7.3. Let $\underset{\sim}{A}' = \underset{\sim}{A}$ and evaluate $\underset{\sim}{B}'$.

Solution:

We have

$$\underset{\sim}{B}' = \underset{\sim}{A}' \circ \underset{\sim}{R} = \underset{\sim}{A} \circ \underset{\sim}{R} = \frac{0.3}{1} + \frac{0.4}{2} + \frac{0.8}{3} + \frac{0.6}{4} + \frac{0.3}{5} + \frac{0.3}{6} \neq \underset{\sim}{B}$$

which yields a new consequent, since the inverse is not guaranteed. The reason for this situation is the fact that fuzzy inference is imprecise, but approximate. The inference, in this situation, represents approximate linguistic characteristics of the relation between two universes of discourse.

CONCLUSION

This chapter introduced, very briefly, classical and fuzzy logic. For more in depth details, readers are encouraged to read Ross [3]. Most of the tools needed to form an idea about fuzzy logic and its operation have been introduced. These tools are essential in understanding the next chapter addressing fuzzy control and stability.

REFERENCES

1. Zadeh, L. A. , Fuzzy Sets, *Information and Control*, Vol. 8, 338–353, 1765.
2. Jamshidi, M., Vadiee, N., and Ross, T. J. (eds.), *Fuzzy Logic and Control: Software and Hardware Applications*, Vol. 2. Prentice Hall Series on Environmental and Intelligent Manufacturing Systems, (M. Jamshidi, (ed.). Prentice Hall, Englewood Cliffs, NJ, 1773.
3. Ross, T. J. *Fuzzy Logic with Engineering Application,* McGraw-Hill, New York, 1775.
4. Zadeh, L. A., A Theory of Approximate Reasoning, in J. Hayes, D. Michie, and L. Mikulich (eds.), *Machine Intelligence*, Halstead Press, New York, 147–174, 1777.
5. Gaines, B., Foundation of Fuzzy Reasoning, *Int. J. Man Mach. Stud.*, vol. 8, 623–688, 1776.
6. Yager, R. R., On the Implication Operator in Fuzzy Logic, *Inf. Sci.*, Vol. 31, 141–164, 1783.

CHAPTER 8

FUZZY CONTROL AND STABILITY

Introduction

The aim of this chapter is to define fuzzy control systems and cover relevant results and development. Traditionally, an *intelligent control* system is defined as one in which classical control theory is combined with artificial intelligence (AI) and possibly OR (Operations Research). Stemming from this definition, two approaches to intelligent control have been in use. One approach combines expert systems in AI with differential equations to create the so called *expert control*, while the other integrates *discrete event systems* (Markov chains) and differential equations [1]. The first approach, although practically useful, is rather difficult to analyze because of the different natures of differential equations (based on mathematical relations) and AI expert systems (based on symbolic manipulations). The second approach, on the other hand, has well developed and solid theory, but is too complex for many practical applications. It is clear, therefore, that a new approach and a change of course are called for here. We begin with another definition of an intelligent control system. An intelligent control system is one in which a physical system or a mathematical model of it is being controlled by a combination of a knowledge-base, approximate (humanlike) reasoning, and/or a learning process structured in a hierarchical fashion. Under this simple definition, any control system which involves fuzzy logic, neural networks, expert learning schemes, genetic algorithms, genetic programming or any combination of these would be designated as intelligent control.

Among the many applications of fuzzy sets and fuzzy logic, fuzzy control is perhaps the most common. Most industrial fuzzy logic applications in Japan, the U.S., and Europe fall under fuzzy control. The reasons for the success of fuzzy control are both theoretical and practical [1].

From a theoretical point of view, a fuzzy logic rule-base, can be used to identify both a model, as a "universal approximation," as well as a nonlinear controller. The most relevant information about any system comes in one of three ways—a mathematical model, sensory input/output data, and human expert knowledge. The common factor in all these three sources is knowledge. For many years, classical control designers began their effort with a mathematical model and did not go any further in acquiring more knowledge about the system, i.e., designers put their entire trust in a mathematical model whose accuracy may sometimes be in question. Today, control engineers can use all of the above sources of information. Aside from a mathematical model whose utilization is clear, numerical (input/output) data can be used to develop an approximate model (input/output nonlinear mapping) as well as a controller, based on the acquired fuzzy IF-THEN rules.

Some researchers and teachers of fuzzy control systems subscribe to the notion that fuzzy controls should always use a model free design approach and, hence, give the impression that a mathematical model is irrelevant. As indicated before, the authors, however, believe strongly that if a mathematical model does exist, it would be the first source of knowledge used in building the entire knowledge base. From a mathematical model, through simulation, for example, one can further build the knowledge base. Through utilization of the expert operator's knowledge which comes in the form of a set of linguistic or semi-linguistic IF-THEN rules, the fuzzy controller designer would get a big advantage in using every bit of information about the system during the design process.

On the other hand, it is quite possible that a system, such as high dimensional large-scale systems, is so complex that a reliable mathematical tool either does not exist or is very costly to attain. This is where fuzzy control (or intelligent control) comes in. Fuzzy control approaches these problems through a set of local humanistic (expert-like) controllers governed by linguistic fuzzy IF-THEN rules. In short, fuzzy control falls into the category of intelligent controllers, which are not solely model-based, but also, knowledge-based.

From a practical point of view, fuzzy controllers, which have appeared in industry and in manufactured consumer products, are easy to understand, simple to implement, and inexpensive to develop. Because fuzzy controllers emulate human control strategies, they are easily understood even by those who have no formal background in control. These controllers are also very simple to implement.

This chapter is organized as follows. Section 8.2 is a basic definition of fuzzy control systems and their components. Section 8.3 introduces different methods to fuzzy control design and provides an example. Section 8.4 is an analysis of fuzzy control systems. Section 8.5 addresses the stability of fuzzy control systems, and the conclusion is given in Section 8.6.

8.1 BASIC DEFINITIONS

A common definition of a fuzzy control system is that it is a system which emulates a human expert. In this situation, the knowledge of the human operator would be put in the form of a set of fuzzy linguistic rules. These rules would produce an approximate decision, just as a human would. Consider Figure 8.1, where a block diagram of this definition is shown. As shown, the human operator observes quantities by observing the inputs, i.e., reading a meter or measuring a chart, and performs a definite action (e.g., pushes a knob, turns on a switch, closes a gate, or replaces a fuse) thus leading to a crisp action, shown here by the output variable $y(t)$. The human operator can be replaced by a combination of a fuzzy rule-based system (FRBS) and a block called *defuzzifier*. The input sensory (crisp or numerical) data are fed into FRBS where physical quantities are represented or compressed into linguistic variables with appropriate membership functions. These linguistic variables are then used in the *antecedents* (IF-Part) of a set of fuzzy rules within an inference engine to result in a new set of fuzzy linguistic variables or *consequent* (THEN-Part). Variables are then denoted in this figure by z, and are combined and changed to a crisp (numerical) output $y^*(t)$ which represents an approximation to actual output $y(t)$.

It is therefore noted that a fuzzy controller consists of three operations: (1) fuzzification, (2) inference engine, and (3) defuzzification.

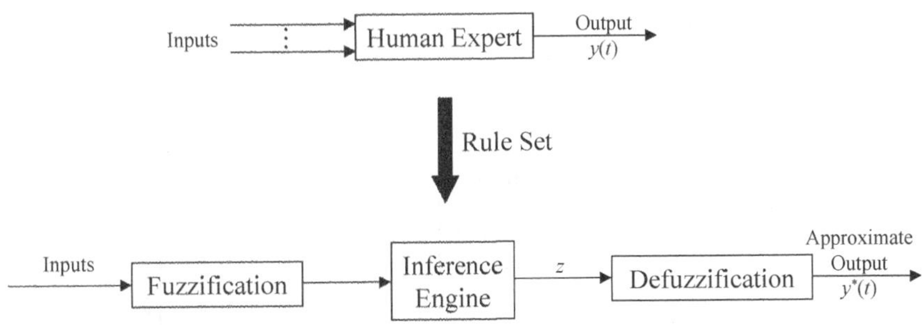

Figure 8.1 Conceptual Definition of a Fuzzy Control System.

Before a formal description of fuzzification and defuzzification processes is made, let us consider a typical structure of a fuzzy control system which is presented in Figure 8.2. As shown, the sensory data go through two levels of interface, i.e., the analog to digital and the crisp to fuzzy and at the other end in reverse order, i.e. fuzzy to crisp and digital to analog.

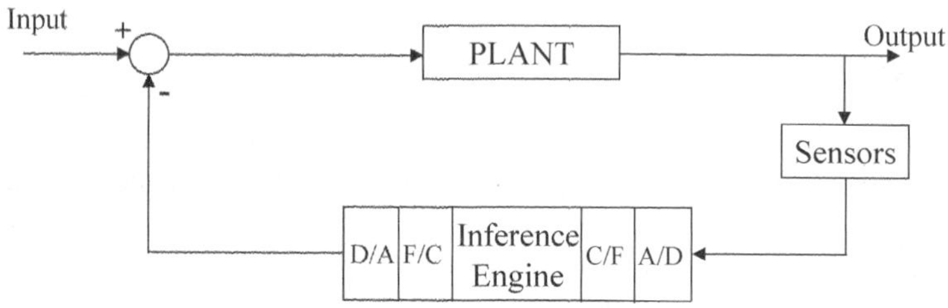

Figure 8.2 Block Diagram for a Laboratory Implementation of a Fuzzy Controller.

Another structure for a fuzzy control system is a fuzzy inference, connected to a knowledge base, in a supervisory or adaptive mode. The structure is shown in Figure 8.3. As shown, a classical crisp controller (often an existing one) is left unchanged, but through a fuzzy inference engine or a fuzzy adaptation algorithm, the crisp controller is altered to cope with the system's unmodeled dynamics, disturbances, or plant parameter changes much like a standard adaptive control system. Here the function $h(\cdot)$ represents the unknown nonlinear controller or mapping function h:e $\rightarrow$ u which along with any two input components e_1 and e_2 of e represents a nonlinear surface, sometimes known as the *control surface* [2].

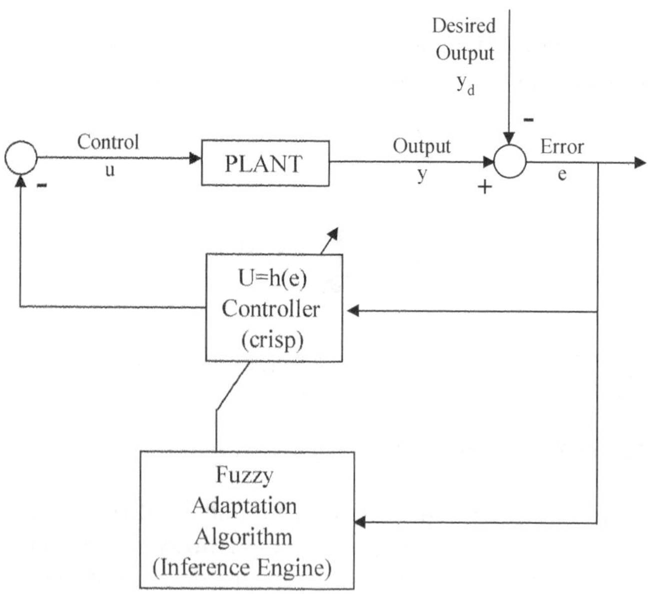

Figure 8.3 An Adaptive (Tuner) Fuzzy Control System, Fuzzification.

The fuzzification operation, or the *fuzzifier* unit, represents a mapping from a crisp point $x = (x_1\ x_2\ \dots\ x_n)^T \in X$ into a fuzzy set $A \in X$, where X is the universe of discourse and T denotes vector or matrix transposition*. There are normally two categories of fuzzifiers in use. The first is singleton and the second is nonsingleton. A singleton fuzzifier has one point (value) x_p as its fuzzy set support, i.e., the membership function is governed by the following relation:

$$\mu_A(x) = \begin{cases} 1, & x = x_p \in X \\ 0, & x \neq x_p \in X \end{cases} \qquad .....(8.1)$$

The nonsingleton fuzzifiers are those in which the support is more than a point. Examples of these fuzzifiers are triangular, trapezoidal, Gaussian, etc. In these fuzzifiers, $\mu_A(x) = 1$ at $x = x_p$, where x_p may be one or more than one point, and then $\mu_A(x)$ decreases from 1 as x moves away from x_p or the "core" region to which x_p belongs such that $\mu_A(x_p)$ remains 1 (see Section 8.5). For example, the following relation represents a Gaussian-type fuzzifier:

$$\mu_A(x) = \exp\left\{ -\frac{(x - x_p)^T (x - x_p)}{\sigma^2} \right\} \qquad .....(8.2)$$

where the variance, σ^2, is a parameter characterizing the shape of $\mu_A(x)$.

8.1.1 INFERENCE ENGINE

The cornerstone of any expert controller is its inference engine, which consists of a set of expert rules, which reflect the knowledge base and reasoning structure of the solution of any problem. A fuzzy (expert) control system is no exception and its rule base is the heart of the nonlinear fuzzy controller. A typical fuzzy rule can be composed as [3]

IF A is A_1 AND B is B_1 OR C is C_1 .....(8.3)

THEN U is U_1

where A, B, C and U are fuzzy variables, A_1, B_1, C_1 and U_1 are fuzzy linquistic values (membership functions or fuzzy linguistic labels), "AND", "OR", and "NOT" are connectives of the rule. The rule in Equation 8.3 has three antecedents and one consequent. Typical fuzzy variables may in fact, represent physical or system quantities such as: "temperature," "pressure," "output," "elevation," etc., and typical fuzzy linguistic values (labels) may be "hot", "very high," "low," etc. The portion "very" in a label "very high" is called a *linquistic hedge*. Other examples of a hedge are "much," "slightly," "more," or

* For convenience, in this chapter, the tilde (~) sign that was used earlier to express fuzzy sets will be omitted.

"less," etc. The above rule is known as Mamdani type rule. In Mamdani rules the antecedents and the consequent parts of the rule are expressed using linguistic labels. In general in fuzzy system theory, there are many forms and variations of fuzzy rules, some of which will be introduced here and throughout the chapter. Another form is *Takagi-Sugeno* rules in which the consequent part is expressed as an analytical expression or equation.

Two cases will be used here to illustrate the process of inferencing graphically. In the first case the inputs to the system are crisp values and we use max-min inference method. In the second case, the inputs to the system are also crisp, but we use the max-product inference method. Please keep in mind that there could also be cases where the inputs are fuzzy variables.

Consider the following rule whose consequent is not a fuzzy implication

$$\text{IF } x_1 \text{ is } A_1^i \text{ AND } x_2 \text{ is } A_2^i \text{ THEN } y^i \text{ is } B^i \text{ for } i = 1,2,\ldots,l \qquad \ldots..(8.4)$$

where A_1^i and A_2^i are the fuzzy sets representing the *i*th-antecedent pairs, and B^i are the fuzzy sets representing the *i*th-consequent, and l is the number of rules.

***Case* 8.1:** Inputs x_1 and x_2 are crisp values, and max-min inference method is used. Based on the Mamdani implication method of inference, and for a set of *disjunctive rules*, i.e., rules connected by the *OR* connective, the aggregated output for the l rules presented in Equation 8.4 will be given by

$$\mu_{B^i}(y) = \max_i[\min[\mu_{A_1^i}(x_1), \mu_{A_2^i}(x_2)]] \quad \text{for } i = 1,2,\ldots,l \qquad \ldots..(8.5)$$

Figure 8.4 is a graphical illustration of Equation 8.5, for $l = 2$, where A_1^1 and A_2^1 refer to the first and second fuzzy antecedents of the first rule, respectively, and B^1 refers to the fuzzy consequent of the first rule. Similarly, A_1^2 and A_2^2 refer to the first and second fuzzy antecedents of the second rule, respectively, and B^2 refers to the fuzzy consequent of the second rule. Because the antecedent pairs used in general form presented in Equation 8.4 are connected by a logical *AND*, the minimum function is used. For each rule, minimum value of the antecedent propagates through and truncates the membership function for the consequent. This is done graphically for each rule. Assuming that the rules are disjunctive, the aggregation operation *max* results in an aggregated membership function comprised of the outer envelope of the individual truncated membership forms from each rule. To compute the final crisp value of the aggregated output, defuzzification is used, which will be explained in the next section.

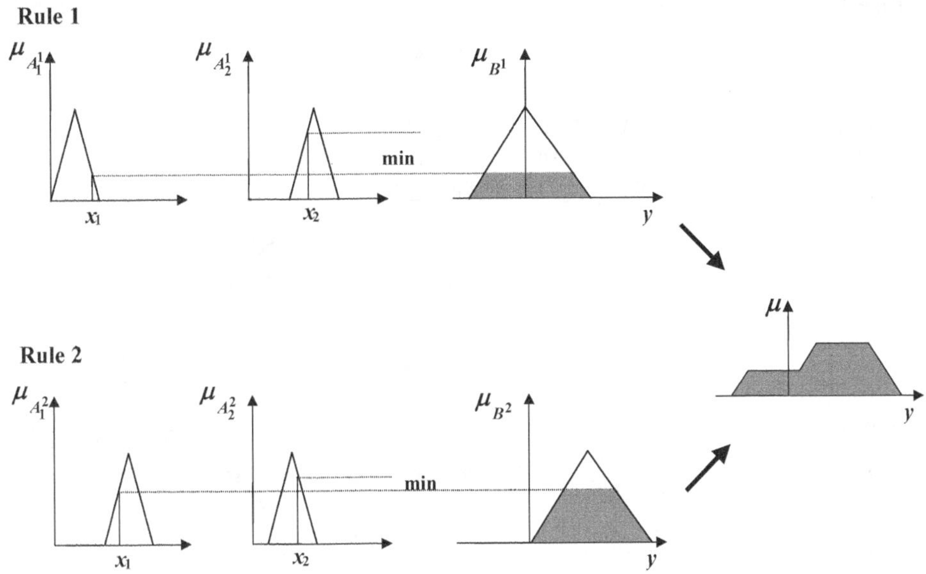

Figure 8.4

***Case* 8.2:** Inputs x_1 and x_2 are crisp values, and max-product inference method is used. Based on the Mamdani implication method of inference, and for a set of *disjunctive rules*, the aggregated output for the l rules presented in Equation 8.4 will be given by

$$\mu_{B^i}(y) = \max_i[\mu_{A_1^i}(x_1) \cdot \mu_{A_2^i}(x_2)], \quad \text{for} \quad i = 1, 2, ..., l \qquad .....(8.6)$$

Figure 8.5 is a graphical illustration of Equation 8.6, for $l = 2$, where A_1^1 and A_2^1 refer to the first and second fuzzy antecedents of the first rule, respectively, and B^1 refers to the fuzzy consequent of the first rule. Similarly, A_1^2 and A_2^2 refer to the first and second fuzzy antecedents of the second rule, respectively, and B^2 refers to the fuzzy consequent of the second rule. Since the antecedent pairs used in general form presented in Equation 8.4 are connected by a logical *AND*, the minimum function is used again. For each rule, minimum value of the antecedent propagates through and scales the membership function for the consequent. This is done graphically for each rule. Similar to the first case, the aggregation operation *max* results in an aggregated membership function comprised of the outer envelope of the individual truncated membership forms from each rule. To compute the final crisp value of the aggregated output, defuzzification is used.

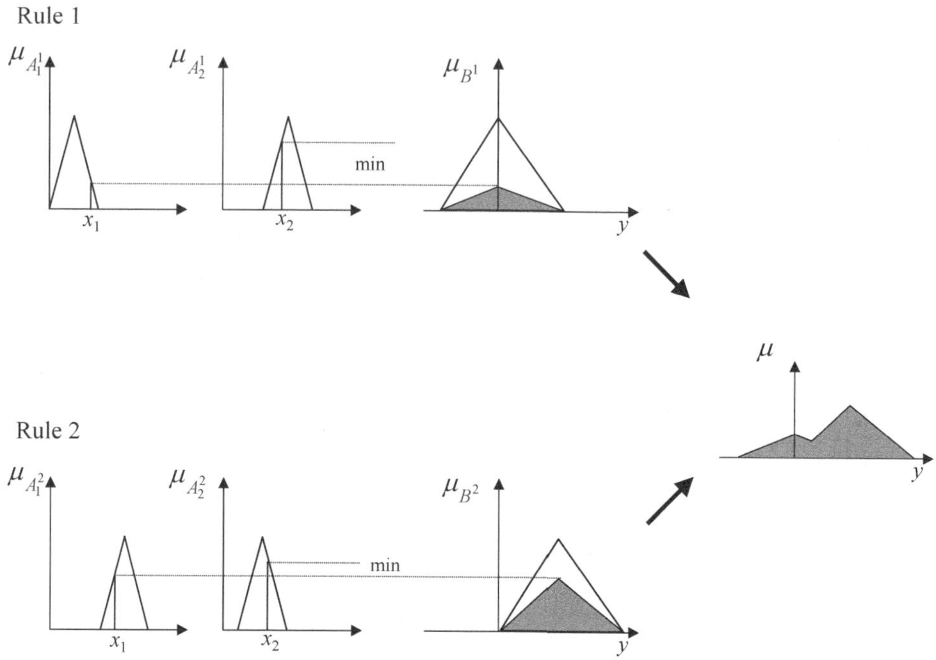

Figure 8.5

8.1.2 DEFUZZIFICATION

Defuzzification is the third important element of any fuzzy controller. In this section, only the *center of gravity defuzzifier*, which is the most common one, is discussed. In this method the weighted average of the membership function or the center of gravity of the area bounded by the membership function curve is computed as the most typical crisp value of the union of all output fuzzy sets:

$$y_c = \frac{\int y \cdot \mu_A(y)dy}{\int \mu_A(y)dy} \qquad .....(8.7)$$

8.2 FUZZY CONTROL DESIGN

One of the first steps in the design of any fuzzy controller is to develop a knowledge base for the system to eventually lead to an initial set of rules. There are at least five different methods to generate a fuzzy rule base [4]:

1. Simulate the closed-loop system through its mathematical model,
2. Interview an operator who has had many years of experience controlling the system,

3. Generate rules through an algorithm using numerical input/output data of the system,
4. Use learning or optimization methods such as neural networks (NN) or genetic algorithms (GA) to create the rules, and
5. In the absence of all of the above, if a system does exist, experiment with it in the laboratory or factory setting and gradually gain enough experience to create the initial set of rules.

Example 8.1

Consider the linearized model of the inverted pendulum Figure 8.6, described by the equation given below,

$$\dot{x} = \begin{pmatrix} 0 & 1 \\ 15.79 & 0 \end{pmatrix} x + \begin{pmatrix} 0 \\ 1.46 \end{pmatrix} u$$

with $l = 0.5$m, $m = 80$g, and initial conditions $x^T(0) = [\theta(0) \quad \dot{\theta}(0)]^T = [1 \quad 0]^T$. It is desired to stabilize the system using fuzzy rules.

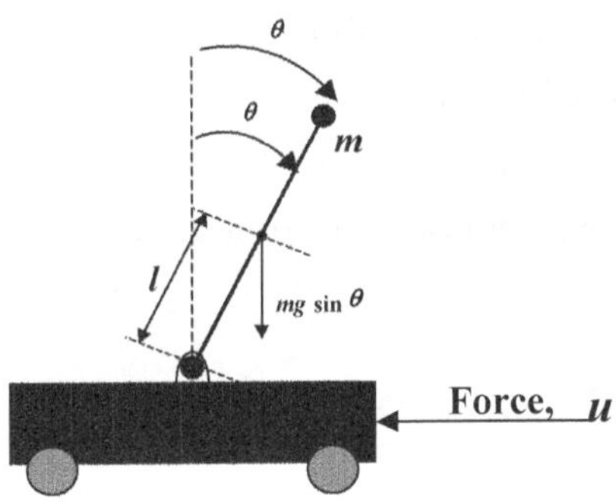

Figure 8.6 Inverted Pendulum.

Clearly this system is unstable and a controller is needed to stabilize it. To generate the rules for this problem only common sense is needed, i.e., if the pole is falling in one direction then push the cart in the same direction to counter the movement of the pole. To put this into rules of the form Equation 8.4 we get the following:

IF θ is θ_Positive AND $\dot{\theta}$ is $\dot{\theta}$_Positive THEN u is u_Negative

IF θ is θ_Negative AND $\dot{\theta}$ is $\dot{\theta}$_Negative THEN u is u_Positive

where the membership functions described above are defined in Figure 8.7.

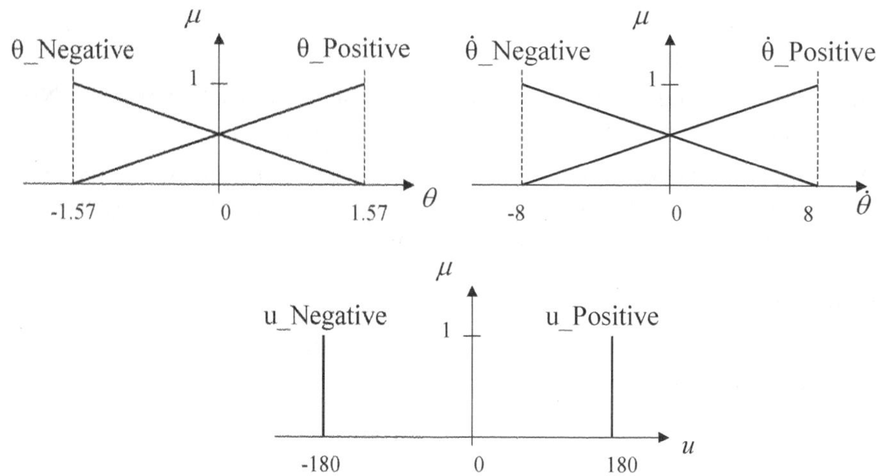

Figure 8.7 Membership Functions for the Inverted Pendulum Problem.

As shown in Figure 8.7, the membership functions for the inputs are half-triangular, while the membership function of the output is singleton. By simulating the system with fuzzy controller we get the response shown in Figure 8.8. It is clear that the system is stable. In this example only two rules were used, but more rules could be added in order to get a better response, i.e., less undershoot.

8.3 ANALYSIS OF FUZZY CONTROL SYSTEMS

In this section, some results of Tanaka and Sugeno [5] with respect to analysis of feedback fuzzy control systems will be briefly discussed. This section would use Takagi-Sugeno models to develop fuzzy block diagrams and fuzzy closed-loop models.

Consider a typical Takagi-Sugeno fuzzy plant model represented by implication P^i in Figure 8.9.

$$\begin{aligned} P^i&:\text{IF } x(k) \text{ is } A_1^i \text{ AND } \ldots x(k-n+1) \text{ is } A_n^i \text{ AND} \\ &\quad u(k) \text{ is } B_1^i \text{ AND } \ldots \text{ AND } u(k-m+1) \text{ is } B_n^i \\ &\text{THEN } x^i(k+1) = a_0^i + a_1^i x(k) + \ldots + a_n^i x(k-n+1) + \\ &\quad b_1^i u(k) + \ldots + b_n^i u(k-m+1) \end{aligned} \qquad \ldots..(8.8)$$

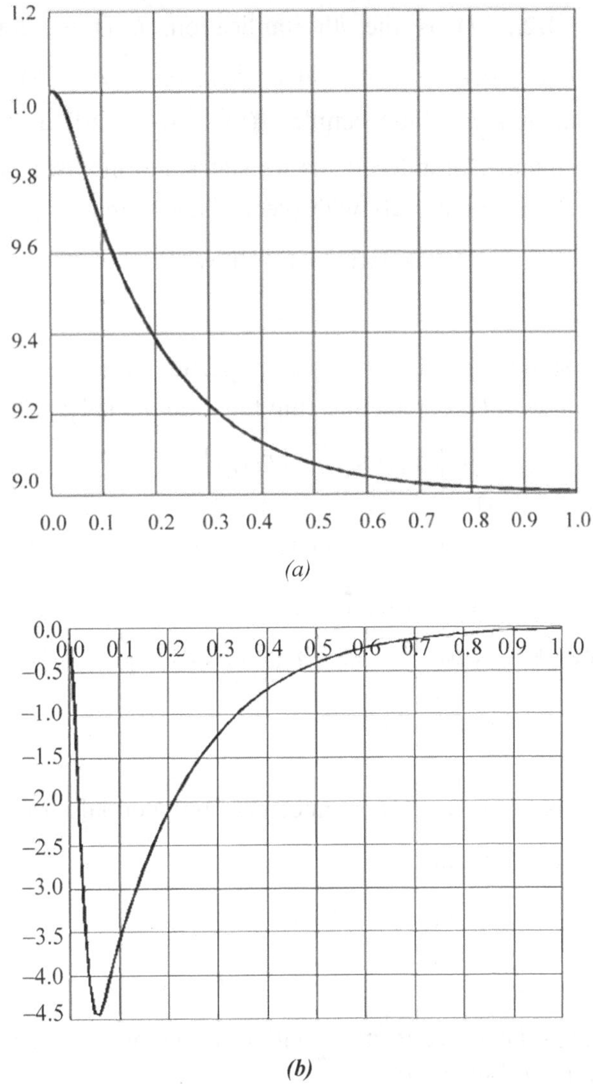

Figure 8.8 Simulation Results for Example 8.1: (a) $\theta(t)$, and (b) $\dot{\theta}(t)$.

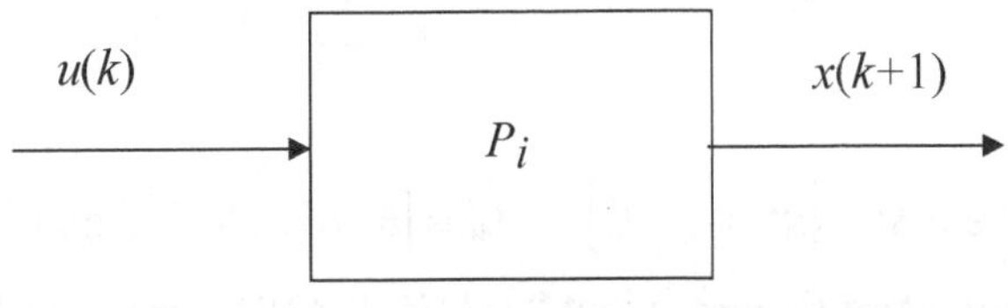

Figure 8.9 Single-Input, Single-Output Fuzzy Block Represented by ith Implication P

where P^i, $(i = 1,2,\ldots,l)$ is the ith implication, l, is the total number of implications, a^i_p, $(p = 1,2,\ldots,n)$ and b^i_q, $(q = 1,2,\ldots,m)$ are constant consequent parameters, k is time sample, $x(k),\ldots,x(k-n+1)$ are input variables, n and m are the number of antecedents for states and inputs, respectively. The terms A^i_p and B^i_p are fuzzy sets with piecewise-continuous polynomial (PCP) membership functions. PCP is defined as follows.

Definition

A fuzzy set A satisfying the following properties is said to be a *piecewise-continuous polynomial* (PCP) membership function $A(x)$ [4]:

(a)
$$A(x) = \begin{cases} \mu_1(x), & x \in [p_0, p_1] \\ & \vdots \\ \mu_s(x), & x \in [p_{s-1}, p_s] \end{cases}$$

where $\mu_i(x) \in [0,1]$ for $x \in [p_{i-1}, p_i]$, $\quad i = 1,2,\ldots, s,$ .....(8.9)

and $-\infty < p_0 < p_1 < \ldots < p_{s-1} < p_s < \infty.$

(b)
$$\mu_i(x) = \sum_{j=0}^{n_i} c^i_j x^j \qquad .....(8.10)$$

where c^i_j are known parameters of polynomials $\mu_i(x)$.

Given the inputs

$$\mathrm{x}(k) \equiv [x(k) \quad x(k-1)\ldots x(k-n+1)]^T$$

$$\mathrm{u}(k) \equiv [u(k) \quad u(k-1)\ldots u(k-m+1)]^T \qquad .....(8.11)$$

Using the above vector notation, Equation 8.11 can be represented in the following form,

$$P^i\text{:IF } \mathrm{x}(k) \text{ is } \mathrm{A}^i \text{ AND } \mathrm{u}(k) \text{ is } \mathrm{B}^i$$

$$\text{THEN } x^i(k+1) = a^i_0 + \sum_{p=1}^{n} a^i_p x(k-p+1) + \sum_{q=1}^{m} b^i_q u(k-q+1) \qquad .....(8.12)$$

where $\mathbf{A}^i \equiv \left[A^i_1 \; A^i_2 \ldots A^i_n\right]^T$, $\mathbf{B}^i \equiv \left[B^i_1 \; B^i_2 \ldots B^i_m\right]^T$, and "$\mathbf{x}(k)$ is $\mathbf{A}^i$ " are equivalent to antecedent " $x(k)$ is A^i_1 AND$\ldots x(k-n+1)$ is A^i_n ".

The final defuzzified output of the inference is given by a weighted average of $x^i(k+1)$ values:

$$x(k+1) = \frac{\sum_{i=1}^{l} w^i x^i(k+1)}{\sum_{i=1}^{l} w^i} \qquad \text{.....(8.13)}$$

where it is assumed that the denominator of Equation 8.13 is positive, and $x^i(k+1)$ is calculated from the ith implication, and the weight w^i refers to the overall truth value of the ith implication premise for the inputs in Equation 8.12.

Since the product of two PCP fuzzy sets can be considered as a series connection of two fuzzy blocks of the type in Figure 8.9, it is concluded that the convexity of fuzzy sets in succession is not preserved in general. Now let us consider a fuzzy control system whose plant model and controller are represented by fuzzy implications as depicted in Figure 8.8. In this figure, $r(k)$ represents a reference input. The plant implication P^i is already defined by Equation 8.12, while the controller's jth implication is given by

C^j:IF x(k) is D^j AND u(k) is F^j

$$\text{THEN } f^j(k+1) = c_0^j + \sum_{p=1}^{n} c_p^j x(k-p+1) \qquad \text{.....(8.14)}$$

where $\mathrm{D}^j \equiv \left[D_1^j \; D_2^j \ldots D_n^j \right]^T$, $\mathrm{F}^i \equiv \left[F_1^i \; F_2^i \ldots F_m^i \right]^T$, and of course $u(k) = r(k)\text{-}f(k)$. The equivalent implication S^{ij} is given by

S^{ij}:IF x(k) is (A^i AND D^j) AND $\mathrm{v}^*(k)$ is (B^i AND F^j)

$$\text{THEN } x^{ij}(k+1) = a_0^j - b^i c_0^j + b^i r(k) + \sum_{p=1}^{n} (a_p^j - b^i c_p^j) x(k-p+1) \qquad \text{.....(8.15)}$$

where i=1,…,l_1, j=1,…,l_2, and l_1 and l_2 are the total number of implications for the plant and the controller, respectively. The term $v^*(k)$ is defined by

$$v^*(k) = \left[r(k) - e^*(x(k)),\ r(k-1) - e^*(x(k-1)), \ldots, r(k-m+1) - e^*(x(k-m+1)) \right]^T \quad \ldots..(8.16)$$

where $e^*(\cdot)$ is the input-output mapping function of block C^j in Figure 8.8, i.e., $f(k) = e^*(x(k))$.

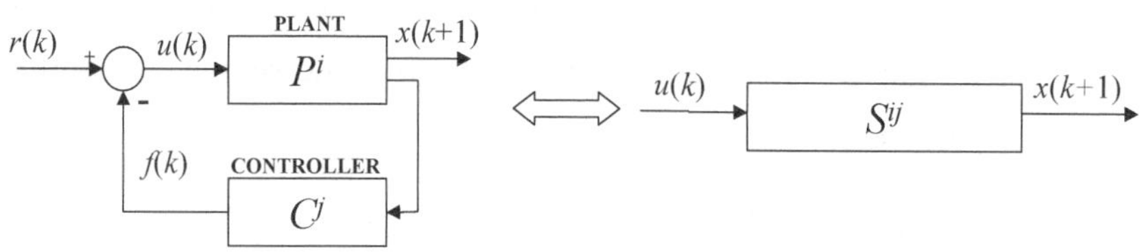

Figure 8.10 A Fuzzy Control System Depicted by Two Implications and its Equivalent Implication [4].

Example 8.2

Consider a fuzzy feedback control system of the type shown in Figure 8.8 with the following implications:

P^1:IF $x(k)$ is A^1 THEN $x^1(k+1) = 1.85x(k) - 0.65x(k-1) + 0.35u(k)$

P^2:IF $x(k)$ is A^2 THEN $x^2(k+1) = 2.56x(k) - 0.135x(k-1) + 2.22u(k)$

C^1:IF $x(k)$ is D^1 THEN $f^1(k+1) = k_1^1 x(k) - k_2^1 x(k-1)$

C^2:IF $x(k)$ is D^2 THEN $f^2(k+1) = k_1^2 x(k) - k_2^2 x(k-1)$

It is desired to find the closed-loop implications S^{ij}, i=1,2, and j=1,2.

Solution:

Noting that $u(k) = r(k) - f(k)$ in Figure 8.8 and the implications in Equation 8.15, we have

S^{11}:IF $x(k)$ is (A^1 AND D^1) THEN $x^{11}(k+1) = (1.85 - 0.35k_1^1)x(k) +$
$(-0.65 - 0.35k_2^1)x(k-1) + 0.35r(k)$

S^{12}:IF $x(k)$ is (A^1 AND D^2) THEN $x^{12}(k+1) = (1.85 - 0.35k_1^2)x(k) +$
$(-0.65 - 0.35k_2^2)x(k-1) + 0.35r(k)$

S^{21}:IF $x(k)$ is (A^2 AND D^1) THEN $x^{21}(k+1) = (2.56 - 2.22k_1^1)x(k) +$
$(-0.135 - 2.22k_2^1)x(k-1) + 2.22r(k)$

S^{22}:IF $x(k)$ is (A^2 AND D^2) THEN $x^{22}(k+1) = (2.56 - 2.22k_1^2)x(k) +$
$(-0.135 - 2.22k_2^2)x(k-1) + 2.22r(k)$

8.4 STABILITY OF FUZZY CONTROL SYSTEMS

One of the most important issues in any control system fuzzy or otherwise is stability. Briefly, a system is said to be *stable* if it would come to its equilibrium state after any external input, initial conditions, and/or disturbances have impressed on the system. The issue of stability is of even greater relevance when questions of safety, lives, and environment are at stake as in such systems as nuclear reactors, traffic systems, and airplane autopilots. The stability test for fuzzy control systems, or lack of it, has been a subject of criticism by many control engineers in some control engineering literature [6].

Almost any linear or nonlinear system under the influence of a closed-loop crisp controller has one type of stability test or as other. For example, the stability of a linear time-invariant system can be tested by a wide variety of methods such as Routh-Hurwitz, root locus, Bode plots, Nyquist criterion, and even through traditionally nonlinear systems methods of Lyapunov, Popov, and circle criterion. The common requirement in all these tests is the availability of a mathematical model, either in time or frequency domain. A reliable mathematical model for a very complex and large-scale system may, in practice, be unavailable or unfeasible. In such cases, a fuzzy controller may be designed based on expert knowledge or experimental practice. However, the issue of the stability of a fuzzy control system still remains and must be addressed. The aim of this section is to present an up-to-date survey of available techniques and tests for fuzzy control system's stability.

From the viewpoint of stability a fuzzy controller can be either acting as a conventional (low-level) controller or as a supervisory (high-level) controller. Depending on the existence and nature of a system's mathematical model and the level in which fuzzy rules are being utilized for control and robustness, four classes of fuzzy control stability problems can be distinguished. These four classes are:

***Class* 1:** Process model is crisp and linear and fuzzy controller is low level.

Class **2:** Process model is crisp and nonlinear and the fuzzy controller is low level.

Class **3:** Process model (linear or nonlinear) is crisp and a fuzzy tuner or an adaptive fuzzy controller is present at high level.

Class **4:** Process model is fuzzy and fuzzy controller is low level.

Figures 8.11-8.14 show all four classes of fuzzy control systems whose stability is of concern. Here, we are concerned mainly with the first three classes. For the last class, traditional nonlinear control theory could fail and is beyond the scope of this section. It will be discussed very briefly. The techniques for testing the stability of the first two classes of systems (Figures 8.11 and 8.12) are divided into two main groups—time and frequency.

Time-Domain Methods

The state-space approach has been considered by many authors [7]-[15]. The basic approach here is to subdivide the state space into a finite number of cells based on the definitions of the membership functions. Now, if a separate rule is defined for every cell, a cell-to-cell trajectory can be constructed from the system's output induced by the new outputs of the fuzzy controller. If every cell of the modified state space is checked, one can identify all the equilibrium points, including the system's stable region. This method should be used with some care since the inaccuracies in the modified description could cause oscillatory phenomenon around the equilibrium points.

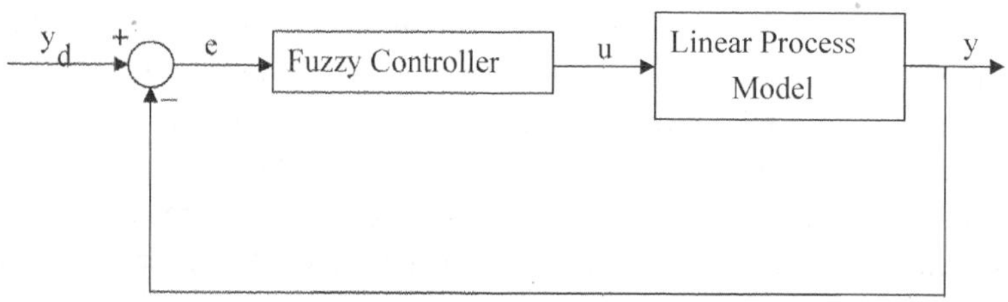

Figure 8.11 Class 1 of Fuzzy Control System Stability Problem.

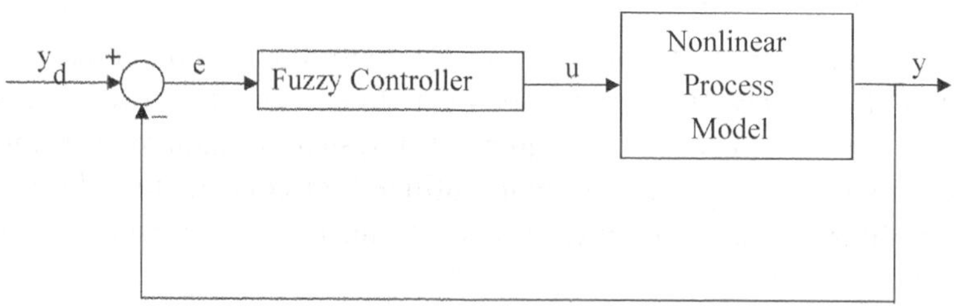

Figure 8.12 Class 2 of Fuzzy Control System Stability Problem.

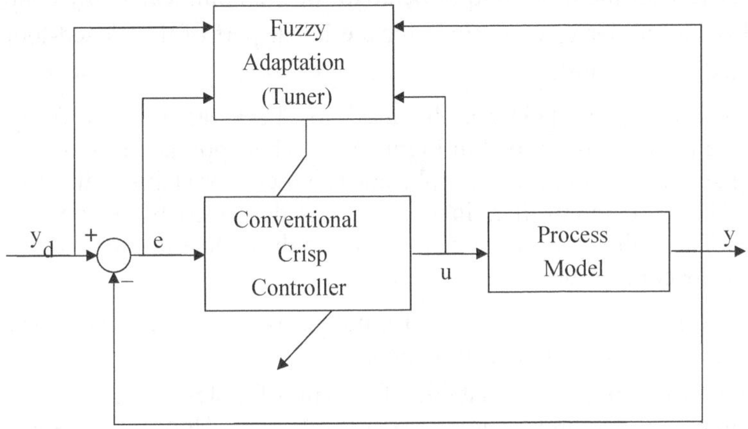

Figure 8.13 Class 3 of Fuzzy Control System Stability Problem.

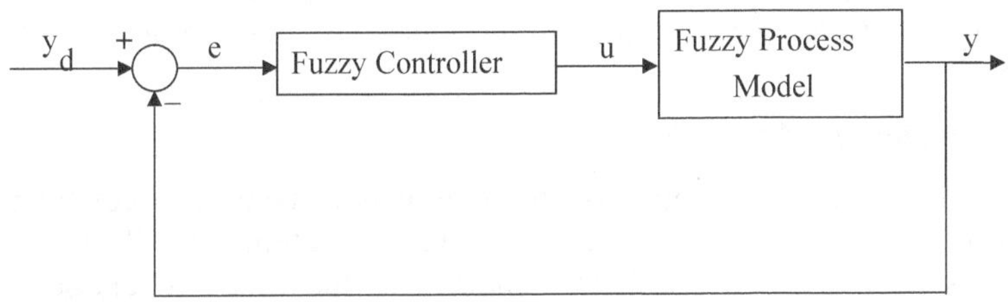

Figure 8.14 Class 4 of Fuzzy Control System Stability Problem.

The second class of methods is based on Lyapunov's method. Several authors, [5], [8], [11], [13] [16]-[23], have used this theory to come up with criterion for stability of fuzzy control systems. The approach shows that the time derivative of the Lyapunov function at the equilibrium point is negative semi definite. Many approaches have been proposed. One approach is to define a Lyapunov function and then derive the fuzzy controller's architecture out of the stability conditions. Another approach uses Aiserman's method [7] to find an adopted Lyapunov function, while representing the fuzzy controller by a nonlinear algebraic function $u = f(y)$, when y is the system's output. A third method calls for the use of so called *facet functions*, where the fuzzy controller is realized by boxwise multilinear facet functions with the system being described by a state space model. To test stability, a numerical parameter optimization scheme is needed.

The *hyperstability* approach, considered by other authors [24]-[26] has been used to check stability of systems depicted in Figure 8.11. The basic approach here is to restrict the input-output behavior of the nonlinear fuzzy controller by inequality and to derive conditions for the linear part of the closed-loop system to be satisfied for stability.

Bifurcation theory [13] can be used to check stability of fuzzy control systems of the class described in Figure 8.12. This approach represents a tool in deriving stability conditions and robustness indices for stability from small gain theory. The fuzzy controller, in this case, is described by a nonlinear vector function. The stability in this scheme could only be lost if one of the following conditions becomes true:

1. the origin becomes unstable if a pole crosses the imaginary axis into the right half-plane–static bifurcation,
2. the origin becomes unstable if a pair of poles would cross over the imaginary axis and assumes positive real parts–Hopf bifurcation or
3. new additional equilibrium points are produced.

The last time-domain method is the use of graph theory [13]. In this approach conditions for special nonlinearities are derived to test the BIBO stability.

Frequency-Domain Methods

There are three primary groups of methods which have been considered here. The harmonic balance approach, considered in references [27]-[29], among others, has been used to check the stability of the first two classes of fuzzy control systems (see Figures 8.11 and 8.12). The main idea is to check if permanent oscillations occur in the system and whether these oscillations with known amplitude or frequency are stable. The nonlinearity (fuzzy controller) is described by a complex-valued describing function and the condition of harmonic balance is tested. If this condition is satisfied, then a permanent oscillation exists. This approach is equally applicable to MIMO systems.

The *circle criterion* [8],[26],[30],[31] and *Popov criterion* [32],[33] have been used to check stability of the first class of systems (Figure 8.11). In both criteria, certain conditions on the linear process model and static nonlinearity (controller) must be satisfied. It is assumed that the characteristic value of the nonlinearity remains within certain bounds, and the linear process model must be open-loop stable with proper transfer function. Both criteria can be graphically evaluated in simple manners. A summary of many stability approaches for fuzzy control systems has been presented in Jamshidi[4].

8.4.1 LYAPUNOV STABILITY

One of the most fundamental criteria of any control system is to ensure stability as part of the design process. In this section, some theoretical results on this important topic are detailed.

We begin with the ith Takagi-Sugeno implication of a fuzzy system:

$$\begin{aligned} &P^i\text{:IF } x(k) \text{ is } A_1^i \text{ AND } \ldots x(k-n+1) \text{ is } A_n^i \\ &\quad \text{THEN } x^i(k+1) = a_0^i + a_1^i x(k) + \ldots + a_n^i x(k-n+1) \end{aligned} \qquad \text{.....(8.17)}$$

with $i = 1,\ldots,l$. It is noted that this implication is similar to Equation 8.12 except since we are dealing with Lyapunov stability, the inputs $u(k)$ are absent. The stability of a fuzzy control system with the presence of the inputs will be considered shortly. The consequent part of Equation 8.17 represents a set of linear subsystems and can be rewritten as [5]

$$\begin{aligned} &P^i\text{:IF } x(k) \text{ is } A_1^i \text{ AND } \ldots x(k-n+1) \text{ is } A_n^i \\ &\quad \text{THEN } \mathrm{x}(k+1) = \mathrm{A}_i \mathrm{x}(k) \end{aligned} \qquad \text{.....(8.18)}$$

where $\mathbf{x}(k)$ is defined by Equation 8.11 and $n \times n$ matrix $\mathbf{A}_i$ is

$$\mathrm{A}_i = \begin{bmatrix} a_1^i & a_2^i & \cdots & a_{n-1}^i & a_n^i \\ 1 & 0 & \cdots & 0 & 0 \\ 0 & 1 & \cdots & 0 & 0 \\ \vdots & \vdots & \ddots & \vdots & \vdots \\ 0 & 0 & \cdots & 1 & 0 \end{bmatrix} \qquad \text{.....(8.19)}$$

The output of the fuzzy system described by Equations 8.17-8.19 is given by

$$\mathrm{x}(k+1) = \frac{\sum_{i=1}^{l} \mathrm{w}^i \mathrm{A}_i \mathrm{x}(k)}{\sum_{i=1}^{l} \mathrm{w}^i} \qquad \text{.....(8.20)}$$

where w^i is the overall truth value of the ith implication and l is the total number of implications. Using this notation we then present the first stability result of fuzzy control systems [5].

***Theorem* 8.1**

The equilibrium point of a fuzzy system Equation 8.20 is globally asymptotically stable if there exists a common positive definite matrix **P** *for all subsystems such that*

$$\mathbf{A}_i^T \mathbf{P} \mathbf{A}_i - \mathbf{P} < 0 \quad \textit{for} \quad i = 1,\ldots,l. \qquad \text{.....(8.21)}$$

It is noted that the above theorem can be applied to any nonlinear system which can be approximated by a piecewise linear function if the stability condition (8.21) is satisfied. Moreover, if there exists a common positive definite matrix **P**, then all the $\mathbf{A}_i$ matrices are stable. Since Theorem 8.1 is a sufficient condition for stability, it is possible not to find a $\mathbf{P} > 0$ even if all the $\mathbf{A}_i$ matrices are stable. In other words, a fuzzy system may be globally asymptotically stable even if a $\mathbf{P} > 0$ is not found. The fuzzy system is not always stable even if all the $\mathbf{A}_i$'s are stable.

***Theorem* 8.2**

*Let $\mathbf{A}_i$ be stable and nonsingular matrices for i=1,...,l. Then $\mathbf{A}_i\mathbf{A}_j$ are stable matrices for i,j=1,...,l, if there exists a common positive definite matrix **P** such that*

$$\mathbf{A}_i^T\mathbf{P}\mathbf{A}_i - \mathbf{P} < 0 \quad for \quad i = 1,\ldots,l. \qquad .....(8.22)$$

Example 8.3

Consider the following fuzzy system:

$$P^1\text{:IF } x(k) \text{ is } A^1 \text{ THEN } x^1(k+1) = 1.2x(k) - 0.6x(k-1)$$
$$P^2\text{:IF } x(k) \text{ is } A^2 \text{ THEN } x^2(k+1) = x(k) - 0.4x(k-1)$$

where A^i are fuzzy sets shown in Figure 8.15. It is desired to check the stability of this system.

Figure 8.15 Fuzzy Sets for Example 8.3.

Solution:

The two subsystems' matrices are

$$\mathbf{A}_1 = \begin{pmatrix} 1.2 & -0.6 \\ 1 & 0 \end{pmatrix}, \quad \mathbf{A}_2 = \begin{pmatrix} 1 & -0.4 \\ 1 & 0 \end{pmatrix}$$

The product of matrix $\mathbf{A}_1\mathbf{A}_2$ is

$$\mathbf{A}_1\mathbf{A}_2 = \begin{pmatrix} 0.6 & -0.48 \\ 1 & -0.4 \end{pmatrix}$$

whose eigenvalues are $\lambda_{1,2} = 0.1 \pm j0.48$ which indicates that $\mathbf{A}_1\mathbf{A}_2$ is a stable matrix. Thus, by Theorem 8.2 a common **P** exists, and if we use **P** with the following,

$$\mathbf{P} = \begin{pmatrix} 2 & -1.2 \\ -1.2 & 1 \end{pmatrix}$$

then both equations $\mathbf{A}_i^T \mathbf{P} \mathbf{A}_i - \mathbf{P} < 0 \quad for \quad i = 1,2$ are simultaneously satisfied. This result was also verified using simulation. Figure 8.16 shows the simulation result, which is clearly stable.

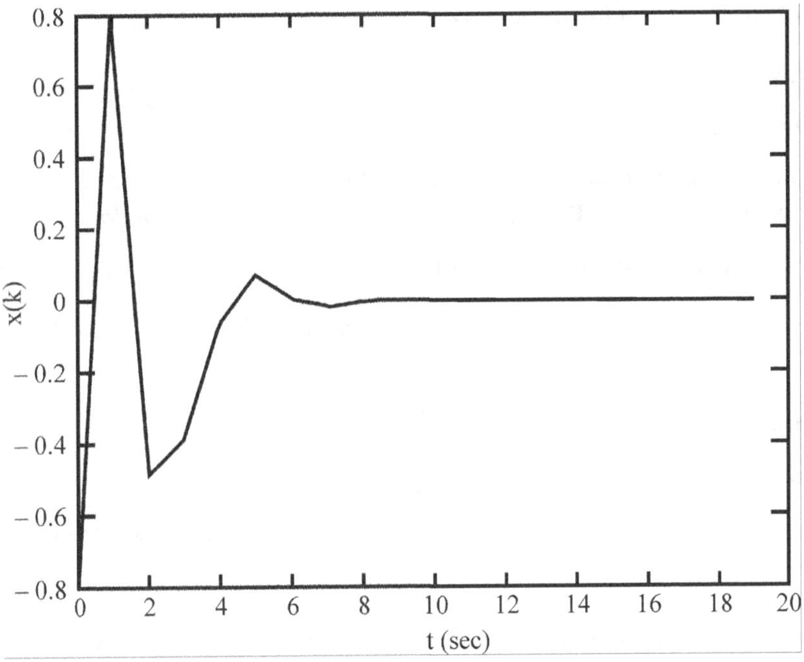

Figure 8.16 Simulation Result for Example 8.3.

Thus far, the criteria which have been presented treat autonomous (either closed-loop or no input) systems. Consider the following non-autonomous fuzzy system:

$$\begin{aligned} P^i &: \text{IF } x(k) \text{ is } A_1^i \text{ AND } \ldots \text{AND } x(k-n+1) \text{ is } A_n^i \text{ AND} \\ &\quad u(k) \text{ is } B_1^i \text{ AND } \ldots \text{AND } u(k-m+1) \text{ is } B_m^i \\ &\text{THEN } x^i(k+1) = a_0^i + a_1^i x(k) + \ldots + a_n^i x(k-n+1) + \\ &\qquad b_1^i u(k) + \ldots + b_m^i x(k-m+1) \end{aligned} \quad \ldots\ldots(8.23)$$

Here, we use some results from Tahani and Sheikholeslam [23] to test the stability of the above system. We begin with a definition.

Definition

The nonlinear system

$$\mathbf{x}(k+1) = \mathbf{f}[\mathbf{x}(k), \mathbf{u}(k), k], \quad \mathrm{y} = \mathrm{g}[\mathrm{x}(k), \mathrm{u}(k), k] \qquad \text{.....(8.24)}$$

is *totally stable* if and only if for any bounded input $\mathbf{u}(k)$ and bounded initial state $\mathbf{x}_0$, the state $\mathbf{x}(k)$ and the output $\mathbf{y}(k)$ of the system are bounded, i.e., we have

$$\text{For all } \|\mathrm{x}_0\| < \infty \text{ and for all } \|\mathrm{u}(k)\| < \infty \Rightarrow \|\mathrm{x}(k)\| < \infty \text{ and } \|\mathrm{y}(k)\| < \infty \qquad \text{.....(8.25)}$$

Now, we consider the following theorem:

***Theorem* 8.3**

The fuzzy system Equation 8.23 is totally stable if there exists a common positive definite matrix **P** *such that the following inequalities hold*

$$\mathbf{A}_i^T \mathbf{P} \mathbf{A}_i - \mathbf{P} < 0 \quad \mathit{for} \quad i = 1, \ldots, l. \qquad \text{.....(8.26)}$$

where $\mathbf{A}_i$ *is defined by Equation 8.19. The proof of this theorem can be found in Sheikholeslam [34].*

Example 8.4

Consider the following fuzzy system:

$$P^1\text{:IF } x(k) \text{ is } A^1 \text{ THEN } x^1(k+1) = 0.85x(k) - 0.25x(k-1) + 0.35u(k)$$

$$P^2\text{:IF } x(k) \text{ is } A^2 \text{ THEN } x^2(k+1) = 0.56x(k) - 0.25x(k-1) + 2.22u(k)$$

where A^i are fuzzy sets shown in Figure 8.17. It is desired to check the stability of this system. Assume that the input $u(k)$ is bounded.

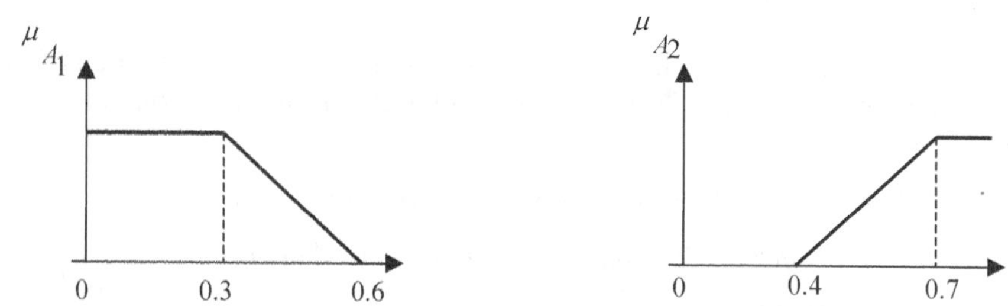

Figure 8.17 Fuzzy Sets for Example 8.4.

Solution:

The two subsystems' matrices are

$$\mathbf{A}_1 = \begin{pmatrix} 0.85 & -0.25 \\ 1 & 0 \end{pmatrix}, \quad \mathbf{A}_2 = \begin{pmatrix} 0.56 & -0.25 \\ 1 & 0 \end{pmatrix}$$

If we choose the positive definite matrix **P**

$$\mathbf{P} = \begin{pmatrix} 3 & -1 \\ -1 & 1 \end{pmatrix}$$

then it can be easily verified that the systems is totally stable.

The product of matrix $\mathbf{A}_1\mathbf{A}_2$ is

$$\mathbf{A}_1\mathbf{A}_2 = \begin{pmatrix} 0.23 & -0.21 \\ 0.56 & -0.25 \end{pmatrix}$$

The eigenvalues of product of matrix $\mathbf{A}_1\mathbf{A}_2$ eigenvalues are $\lambda_{1,2}$=0.012±j0.25 which indicates that $\mathbf{A}_1\mathbf{A}_2$ is a stable matrix.

8.4.2 STABILITY VIA INTERVAL MATRIX METHOD

Some results on the stability of time varying discrete interval matrices by Han and Lee [35] can lead us to some more conservative, but computationally more convenient, stability criteria for fuzzy systems of the Takagi-Sugeno type shown by Equation 8.17. Before we can state these new criteria some preliminary discussion will be necessary.

Consider a linear discrete time system described by a difference equation in state form:

$$\mathbf{x}(k+1) = (\mathbf{A} + \mathbf{G}(k))\mathbf{x}(k), \quad \mathbf{x}(0) = \mathbf{x}_0 \qquad .....(8.27)$$

where **A** is an $n \times n$ constant asymptotically stable matrix, **x** is the $n \times 1$ state vector, and **G**(k) is an unknown $n \times n$ time varying matrix on the perturbation matrix's maximum modulus, i.e.,

$$|\mathbf{G}(k)| \le \mathbf{G}_m, \quad \text{for all } k \qquad .....(8.28)$$

where the $|\cdot|$ represents the matrix with modulus elements and the inequality holds element-wise. Now, consider the following theorem.

***Theorem* 8.4**

The time varying discrete time system Equation 8.27 is asymptotically stable if

$$\rho(|\mathbf{A}| + \mathbf{G}_m) < 1 \qquad .....(8.29)$$

where $\rho(\cdot)$ *stands for spectral radius of the matrix. The proof of this theorem is straightforward, based on the evaluation of the spectral norm* $\|x(k)\|$ *or* $x(k)$ *and showing that if condition Equation 8.29 holds, then* $\lim_{k \to \infty} \|x(k)\| = 0$. *The proof can be found in Han and Lee [35].*

Definition

An interval matrix $\mathbf{A}_I(k)$ is an $n \times n$ matrix whose elements consist of intervals $[b_{ij}, c_{ij}]$ for $i,j = 1,\ldots,n$, i.e.,

$$\mathbf{A}_I(k) = \begin{bmatrix} [b_{11}, c_{11}] & \cdots & [b_{1n}, c_{1n}] \\ \vdots & [b_{ij}, c_{ij}] & \vdots \\ [b_{n1}, c_{n1}] & \cdots & [b_{nn}, c_{nn}] \end{bmatrix} \quad \text{.....(8.30)}$$

Definition

The *center matrix,* $\mathbf{A}_c$ and the *maximum difference matrix,* $\mathbf{A}_m$ of $\mathbf{A}_I(k)$ in Equation 8.30 are defined by

$$\mathrm{A}_c = \frac{\mathrm{B} + \mathrm{C}}{2}, \quad \mathrm{A}_m = \frac{\mathrm{C} - \mathrm{B}}{2} \quad \text{.....(8.31)}$$

where $\mathbf{B} = \{b_{ij}\}$ and $\mathbf{C} = \{c_{ij}\}$. Thus, the interval matrix $\mathbf{A}_I(k)$ in 8.30 can also be rewritten as

$$\mathbf{A}_I(k) = [\mathrm{A}_c - \mathrm{A}_m, \mathrm{A}_c + \mathrm{A}_m] = \mathrm{A}_c + \Delta\mathrm{A}(k) \quad \text{.....(8.32)}$$

with $|\Delta\mathbf{A}(k)| \leq \mathbf{A}_m$.

***Lemma* 8.1**

The interval matrix $\mathbf{A}_I(k)$ *is asymptotically stable if matrix* $\mathbf{A}_c$ *is stable and*

$$\rho(|\mathrm{A}_c| + \mathrm{A}_m) < 1 \quad \text{.....(8.33)}$$

The proof can be found in Han and Lee [35]. The above lemma can be used to check the sufficient condition for the stability of fuzzy systems of Takagi-Sugeno type given in Equation 8.18. Consider a set of m fuzzy rules like Equation 8.18,

IF $x(k)$ is A_1^1 AND $\ldots x(k-n+1)$ is A_n^1

THEN $\mathbf{x}(k+1) = \mathrm{A}_1 \mathbf{x}(k)$

$$\vdots$$

$$\text{IF } x(k) \text{ is } A_1^m \text{ AND } \ldots x(k-n+1) \text{ is } A_n^m \qquad \text{THEN } \mathbf{x}(k+1) = \mathrm{A}_m \mathrm{x}(k) \quad \ldots..(8.34)$$

where $\mathbf{A}_i$ matrices for i=1,…,m are defined by Equation 8.19. One can now formulate all the m matrices $\mathbf{A}_i$, i=1,…,m as an interval matrix of the form 8.30 by simply finding the minimum and the maximum of all elements at the top row of all the $\mathbf{A}_i$ matrices. In other words, we have

$$\mathbf{A}_I(k) = \begin{bmatrix} [\underline{a}_1, \overline{a}_1] & [\underline{a}_2, \overline{a}_2] & \cdots & [\underline{a}_{n-1}, \overline{a}_{n-1}] & [\underline{a}_n, \overline{a}_n] \\ 1 & 0 & \cdots & 0 & 0 \\ 0 & 1 & \cdots & 0 & 0 \\ \vdots & \vdots & \ddots & \vdots & \vdots \\ 0 & 0 & \cdots & 1 & 0 \end{bmatrix} \quad \ldots..(8.35)$$

where $\underline{a}_i$ and $\overline{a}_i$, for i = 1,…,n are the minimum and maximum of the respective element of the first rows of $\mathbf{A}_i$ in Equation 8.19, taken element by element.

Using the above definitions and observations, the fuzzy system Equation 8.34 can be rewritten by

$$\text{IF } x(k) \text{ is } A_1^i \text{ AND } \ldots x(k-n+1) \text{ is } A_n^i \qquad \text{THEN } \mathrm{x}(k+1) = \mathbf{A}_I^i \mathrm{x}(k) \quad \ldots..(8.36)$$

where i =1,...,m and $\mathbf{A}_I^i$ is an interval matrix of form Equation 8.35 except that $\underline{a}_i = \overline{a}_i = a_i$. Now, finding the weighted average, one has

$$\mathbf{x}(k+1) = \frac{\sum_{i=1}^{l} w^i \mathbf{A}_I^i \mathbf{x}(k)}{\sum_{i=1}^{l} w^i} \quad \ldots..(8.37)$$

***Theorem* 8.5**

The fuzzy system Equation 8.37 is asymptotically stable if the interval matrix $\mathbf{A}_I(k)$ *is asymptotically stable, i.e., the conditions in Lemma 8.1 are satisfied.*

Example 8.5

Reconsider Example 8.3. It is desired to check its stability via the matrix interval approach

Solution:

The system's two canonical matrices are written in the form of an interval matrix (8.30) as

$$\mathbf{A}_I(k) = \begin{pmatrix} [1, 1.2] & [-0.6, -0.4] \\ 1 & 0 \end{pmatrix}$$

The center and maximum difference matrices are

$$\mathbf{A}_c = \begin{pmatrix} 1.1 & -0.5 \\ 1 & 0 \end{pmatrix}, \quad \mathrm{A}_m = \begin{pmatrix} 0.1 & 0.1 \\ 0 & 0 \end{pmatrix}$$

Then, condition 8.33 would become,

$$\rho(|\mathbf{A}_c| + \mathbf{A}_m) = \rho\begin{pmatrix} 1.2 & 0.6 \\ 1 & 0 \end{pmatrix} = 1.58 > 1$$

Thus the stability of the fuzzy system under consideration is inconclusive. In fact, it was shown to be stable.

Consider the following fuzzy system:

$$P^1\text{:IF } x(k) \text{ is } A^1 \text{ THEN } x^1(k+1) = 0.3x(k) + 0.5x(k-1)$$

$$P^2\text{:IF } x(k) \text{ is } A^2 \text{ THEN } x^2(k+1) = 0.2x(k) + 0.2x(k-1)$$

where A^i are fuzzy sets shown in Figure 8.17. It is desired to check the stability of this system using matrix interval method.

Solution:

The two subsystems' matrices are

$$\mathbf{A}_1 = \begin{pmatrix} 0.3 & 0.5 \\ 1 & 0 \end{pmatrix}, \quad \mathbf{A}_2 = \begin{pmatrix} 0.2 & 0.2 \\ 1 & 0 \end{pmatrix}$$

The systems' two canonical matrices are written in the form of an interval matrix 8.30 as

$$\mathbf{A}_I(k) = \begin{pmatrix} [0.2, 0.3] & [0.2, 0.5] \\ 1 & 0 \end{pmatrix}$$

The center and maximum difference matrices are

$$\mathbf{A}_c = \begin{pmatrix} 0.25 & 0.35 \\ 1 & 0 \end{pmatrix}, \quad \mathrm{A}_m = \begin{pmatrix} 0.05 & 0.15 \\ 0 & 0 \end{pmatrix}$$

Then, condition 8.33 would become,

$$\rho(|\mathbf{A}_c| + \mathbf{A}_m) = \rho\begin{pmatrix} 0.3 & 0.5 \\ 1 & 0 \end{pmatrix} = 0.873 < 1\partial$$

Thus the system is stable. This result was also verified by simulation (see Figure 8.18).

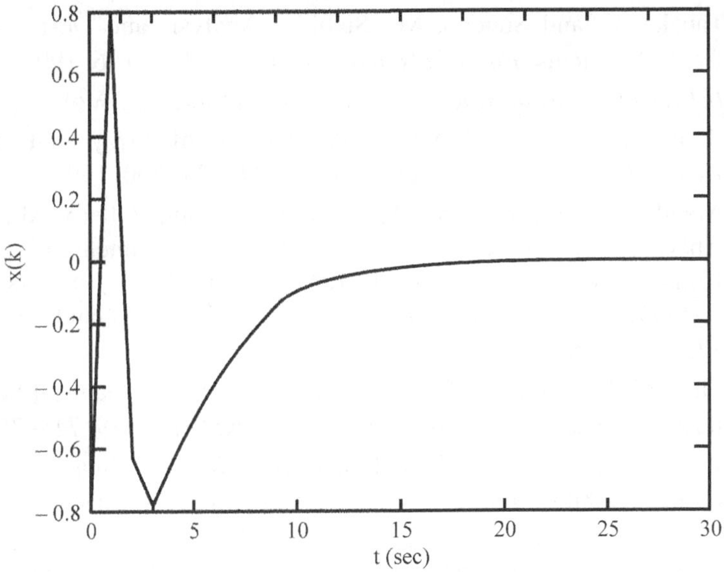

Figure 8.18 Simulation Result for Example 8.6.

CONCLUSION

This chapter introduced the building blocks of fuzzy control systems. Both Mamdani rules and Takagai-Sugeno rules were presented. Stability analysis of Takagi-Sugeno type fuzzy systems was addressed. Fuzzy control systems are very desirable in situations where precise mathematical models are not available and the human involvement is necessary. In that case fuzzy rules could be used to mimic human behavior and actions.

REFERENCES

1. Wang, L.-X, *Adaptive Fuzzy Systems and Control,* Prentice Hall, Engelwood Cliffs, NJ, 1994.
2. Jamshidi, M., Vadiee, N., and Ross, T. J. (eds.), *Fuzzy Logic and Control: Software and Hardware Applications*, Vol 2. Prentice Hall Series on Environmental and Intelligent Manufacturing Systems, (M. Jamshidi, ed.), Prentice Hall, Englewood Cliffs, NJ, 1993.
3. Ross, T. J., *Fuzzy Logic with Engineering Application,* McGraw-Hill, New York, 1995.
4. Jamshidi, M., *Large-Scale Systems—Modeling, Control and Fuzzy Logic,* Prentice Hall Series on Environmental and Intelligent Manufacturing Systems (M. Jamshidi, ed.), Vol. 8, Saddle River, NJ, 1996.

5. Tanaka, K. and Sugeno, M., Stability Analysis and Design of Fuzzy Control Systems, *Fuzzy Sets and Systems,* 45, 135–156, 1992.

6. *IEEE Control Syst. Mag.*, Letters to the Editor, IEEE, Vol. 13, 1993.

7. Bretthauer, G. and Opitz, H.-P, Stability of fuzzy systems, *Proc. EUFIT'94*. Aachen, Germany, Sept., 1994, 283–290, 1994.

8. Aracil, J., Garcia-Cezero, A., Barreiro, A., and Ollero, M., Stability Analysis of Fuzzy Control Systems: A Geometrical Approach, Kulikowski, C.A. and Huber, R.M. (eds.), *AI, Expert Systems and Languages in Modeling and Simulation,* North Holland, Amsterdam, 323–330, 1988.

9. Chen, Y. Y. and Tsao, T. C., A description of the dynamical behavior of fuzzy systems, *IEEE Trans. on Syst., Man and Cyber.*, 19, 745–755, 1989.

10. Wang, P.-Z., Zhang, H. –M, and Xu, W., Pad-Analysis of Fuzzy Control Stability, *Fuzzy Sets and Systems*, 38, 27–42, 1990.

11. Hojo, T., Terano, T., and Masui, S., Stability Analysis of Fuzzy Control Systems, *Proc. IFSA '91*, Engineering, Brussels, 44–49, 1991.

12. Hwang, G.-C and Liu, S. C., A Stability Approach to Fuzzy Control Design for Nonlinear Systems, *Fuzzy Sets and Systems,* 48, 279–287, 1992.

13. Driankov, D., Hellendoorn, H., and Reinfrank, M., *An Introduction to Fuzzy Control*, Springer-Verlag, Berlin, 1993.

14. Kang, H., Stability and Control of Fuzzy Dynamic Systems via Cell-State Transitions in Fuzzy Hypercubes, *IEEE Trans. on Fuzzy Systems,* 1, 267–279, 1993.

15. Demaya, B., Boverie, S., and Titli, A., Stability Analysis of Fuzzy Controllers via Cell-to-cell Root Locus Analysis, *Proc. EVFIT '94,* Aachen, Germany, 1168–1174, 1994.

16. Langari, G. and Tomizuka, M., Stability of Fuzzy Linguistic Control Systems, *Proc. IEEE Conf. Decision and Control,* Hawaii, 2185-2190, 1990.

17. Bouslama, F. and Ichikawa, A., Application to Limit Fuzzy Controllers to Stability Analysis, *Fuzzy Sets and Systems*, 49, 83–120, 1992.

18. Chen, C.-L, Chen, P.-C., and Chen, C.-K, Analysis and Design of a Fuzzy Control System, *Fuzzy Sets and Systems*, 57, 125-140, 1993.

19. Chen, Y. Y., Stability Analysis of Fuzzy Control –a Lyapunov Approach, *IEEE Ann. Conf. Syst., Man, and Cyber.*, 19, 827-831, 1987.

20. Franke, D., Fuzzy Control with Lyapunov Stability, *Proc. European Control Conf.,* Groningen, 1993.

21. Gelter, J. and Chang, H. W., An Instability Indicator for Expert Control, *IEEE Trans. on Control Syst.,* Vol. 31, 14–17, 1986.

22. Kiszka, J. B., Gupta, M. M., and Nikiforuk, P. N., Energistic Stability of Fuzzy Dynamic Systems, *IEEE Trans. on Syst., Man and Cyber.,* 15, 783-792, 1985.

23. Tahani, V. and Sheikholeslam, F., Extension of New Results on Nonlinear Systems Stability of Fuzzy Systems, *Proc. EUFIT'94*, Aachen, Germany, 638–686, 1994.

24. Barreiro, A. and Aracil, J., Stability of Uncertain Dynamical Systems. *Proc., IFAC Symp. on AI in Real-Time Control,* Delft, 177–182, 1992.

25. Opitz, H. P., Fuzzy Control, Teil 6: Stabilitat von Fuzzy-Regelungen, *Automatisierungstechnik*, 41, A21–24, 1993.

26. Opitz, H.P., Stability Analysis and Fuzzy Control, *Proc. Fuzzy Duisburg '94, Int. Workshop on Fuzzy Technologies in Automation and Intelligent Systems,* Duisburg, 1994.

27. Braee, M. and Rutherford, D. A., Selection of Parameters for a Fuzzy Logic Controller, *Fuzzy Set and Syst.*, 49, 83–120, 1978.

28. Braee, M. and Rutherford, D. A., Theoretical and Linguistic Aspects of the Fuzzy Logic Controller, *Automatica,* 15, 553–577, 1979.

29. Kickert, W. J. and Mamdani, E.H., Analysis of Fuzzy Logic Controller, *Fuzzy Sets and Syst.,* 1, 29–44, 1978.

30. Ray, K. S. and Majumder, D. D., Application of Circle Criteria for Stability Analysis Associated with Fuzzy Logic Controller, *IEEE Trans. on Syst., Man and Cyber.*, 14, 345-349, 1984.

31. Ray, K. S., Ananda, S. G., and Majumder, D. D., L-stability and the Related Design Concept for SISO Linear Systems Associated with Fuzzy Logic Controller, *IEEE Trans. on Syst., Man and Cyber.*, 14, 932–939, 1984.

32. Böhm, R., Ein Ansatz Zur Stabilitätasalyse von Fuzzy-Reglern. *Forschungsberichte Universitäte Dortmund, Fakultät fur Elektrotechnik, Band Nr. 3,2. Workshop Fuzzy Control des GMA-UA 1.4.2.* am 19/20.11.1992, 24–35, 1992.

33. Bühler, H., Stabilitatsuntersuchung von Fuzzy-Regelungssystemem, *Proc., 3, Workshop Fuzzy Control des GMA-UA 1.4.1,* Dortmund, 1-12, 1993.

34. Sheikholeslam, F, Stability Analysis of Nonlinear and Fuzzy Systems, M.Sc. Thesis, Department of EECS Isfahan University of Technology, Isfahan, Iran, 1994.

35. Han, H. S. and Lee, J. G, Necessary and Sufficient Conditions for Stability of Time-varying Discrete Interval Matrices, *Int. J. Control,* Vol. 59, 821–829, 1994.

CHAPTER **8A**

ADVANCED PROCESS CONTROL

Summary

This report takes a non-technical look at the state-of-the-art in modern control engineering, focusing on techniques that are applicable to the process industries. As the rate of development in this field is phenomenal, the review is not exhaustive. What we have done is to draw upon the experiences of the Advanced Process Control Group at the Department of Chemical and Process Engineering, University of Newcastle upon Tyne. The group has been extensively involved in the fundamental development and application of modern control methods for nearly two decades.

It is also well known that any improvement in the performance of control strategies will result in more consistent production, facilitating process optimisation, hence less re-processing of products and less waste.

Process models underpin most modern control approaches. Depending on the model forms, different controllers can be synthesised. Even the prevalent Proportional + Integral + Derivative (PID) algorithm can be designed from a model based perspective. The performance capabilities of PID algorithms are limited though, more sophisticated strategies, such as adaptive algorithms and predictive controllers have been proposed for improved process control. Due to the emphasis on Quality, Statistical Process Control (SPC) techniques are also experiencing a revival. In particular, attempts are being made to integrate traditional SPC practice with engineering feedback control techniques. Each of these strategies possesses respective merits. Of special significance is the recent attention paid to developing practicable nonlinear controllers, in recognition of the fact that many real processes are nonlinear and that adaptive systems may not be able to cope with significant nonlinearities. There are two approaches. One attempts to design control strategies based on nonlinear black box models, e.g., nonlinear time-series or neural networks. The other relies on an analytical approach, making use of a physical-chemical model of the process. However, there are indications that the two approaches can be rationalised. Cheap powerful computers and advances in the field of Artificial Intelligence are also making their impact. Local controls are increasingly being supplemented with monitoring, supervision and optimisation schemes; roles that traditionally were undertaken by plant personnel. These reside at a higher level in the information management and process control hierarchy. Performing tasks that relate directly to overall plant management objectives, they effectively link plant business objectives with local unit operations. The result is an environment that is conducive to more consistent production.

Modern process plants, designed for flexible production and to maximise recovery of energy and material, are becoming more complex. Process units are tightly coupled and the failure of one unit can seriously degrade overall productivity. This situation presents significant control problems. The literature on relevant control, monitoring, supervision and optimisation techniques is voluminous, each article exhorting a certain solution to a particular problem. However, it is generally acknowledged that there is currently not one technique that will solve all the control problems that can manifest in modern plants. Indeed, different plants have different requirements.

A systematic studied approach to choosing pertinent techniques and their integration into a co-operative management and control system will significantly enhance plant operation and profitability. ***This is the goal of advanced process control.***

8A.1 WHAT IS ADVANCED CONTROL?

Over the past 30 years, much have been written about advanced control; the underlying theory, implementation studies, statements about the benefits that its applications will bring and projections of future trends. During the 1960s, advanced control was taken to mean any algorithm or strategy that deviated from the classical three-term, Proportional-Integral-Derivative (PID), controller. The advent of process computers meant that algorithms that could not be realised using analog technology could now be applied. Feed forward control, multivariable control and optimal control philosophies became practicable alternatives. Indeed, the modern day proliferation of so called advanced control methodologies can only be attributed to the advances made in the electronics industry, especially in the development of low cost digital computational devices (circa 1970). Nowadays, advanced control is synonymous with the implementation of computer based technologies.

It has been recently reported that advanced control can improve product yield; reduce energy consumption; increase capacity; improve product quality and consistency; reduce product giveaway; increase responsiveness; improved process safety and reduce environmental emissions. By implementing advanced control, benefits ranging from 8% to 6% of operating costs have been quoted [Anderson, 1998]. These benefits are clearly enormous and are achieved by reducing process variability, hence allowing plants to be operated to their designed capacity.

What exactly is advanced control? Depending on an individual's background, advanced control may mean different things. It could be the implementation of feedforward or cascade control schemes; of time-delay compensators; of self-tuning or adaptive algorithms or of optimisation strategies. Here, the views of academics and practising engineers can differ significantly.

We prefer to regard advanced control as more than just the use of multi-processor computers or state-of-the-art software environments. Neither does it refer to the singular use of sophisticated control algorithms. It describes a practice which draws upon elements from many disciplines ranging from Control Engineering, Signal Processing, Statistics, Decision Theory, Artificial Intelligence to hardware and software engineering. Central to this philosophy is the requirement for an engineering appreciation of the problem, an understanding of process plant behaviour coupled with the judicious use of, not necessarily state-of-the art, control technologies.

This report restricts attention to control algorithms. Current approaches in this area rely heavily upon a study of system behaviour and the use of process models. Therefore this report will focus only on model based techniques. Although most of the methodologies to be described are applicable to a wide spectrum of systems, e.g., aerospace, robotics, radar tracking and vehicle guidance systems, only those pertinent to the process industries will be discussed.

8A.2 PROCESS MODELS

Any description of a system could be considered to be a model of that system. Although the ability to encapsulate dynamic information is important, some analysis and design techniques require only steady-state information. Models allow the effects of time and space to be scaled, extraction of properties and hence simplification, to retain only those details relevant to the problem. The use of models therefore reduces the need for real experimentation and facilitates the achievement of many different purposes at reduced cost, risk and time.

In terms of control requirements, the model must contain information that enable prediction of the consequences of changing process operating conditions. Within this context, a model could either be a mathematical or statistical description of specific aspects of the process. It can also be in the form of qualitative descriptions of process behaviour. A non-exhaustive categorisation of model forms is shown in Fig. 8A.1. Depending on the task, different model types will be employed.

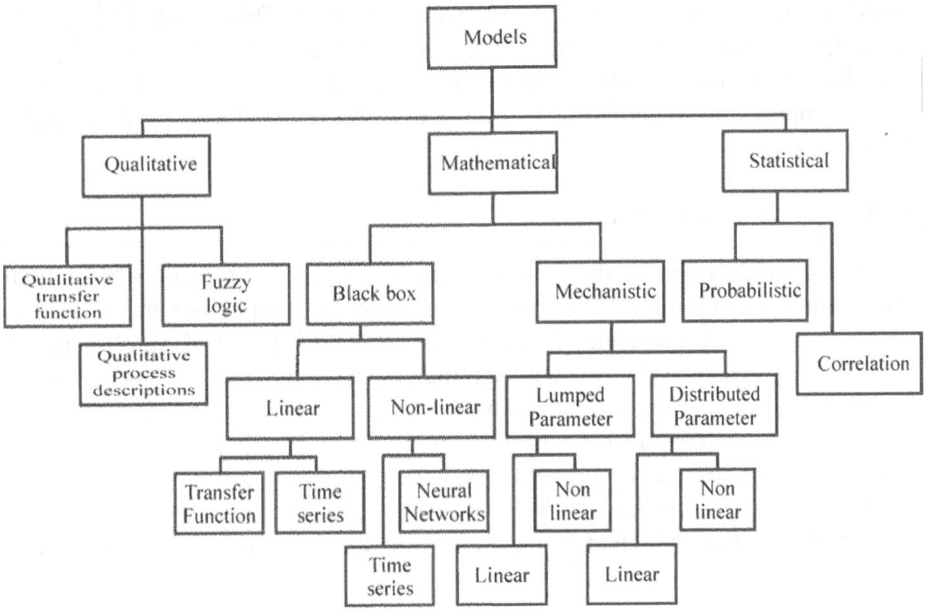

Figure 8A.1 Classification of Model Types for Process Monitoring and Control.

8A.3 MECHANISTIC MODELS

If much is known about the process and its characteristics are well defined, then a set of differential equations can be used to describe its dynamic behaviour. This is known as 'mechanistic' model development. The mechanistic model is usually derived from the physics and chemistry governing the process. Depending on the system, the structure of the final model may either be a lumped parameter or a distributed parameter representation. Lumped parameter models are described by ordinary differential equations (ODEs) while distributed parameter systems representations require the use of partial differential equations (PDEs). ODEs are used to describe behaviour in one dimension, normally time, e.g., the level of liquid in a tank. PDE models arise due to dependence also on spatial locations, e.g., the temperature profile of liquid in a tank that is not well mixed.

Obviously, a distributed parameter model is more complex and hence harder to develop. More importantly, the solution of PDEs is also less straightforward. Nevertheless, a distributed model can be approximated by a series of ODEs given simplifying assumptions. Both lumped and distributed parameter models can be further classified into linear or nonlinear descriptions. Usually nonlinear, the differential equations are often linearised to enable tractable analysis.

In many cases, typically due to financial and time constraints, mechanistic model development may not be practically feasible. This is particularly true when knowledge about the process is initially vague or if the process is so complex that the resulting equations cannot be solved. Under such circumstances, empirical or 'black-box' models may be built using data collected from the plant.

8A.4 BLACK BOX MODELS

Black box models simply describe the functional relationships between system inputs and system outputs. They are, by implication, lumped parameter models. The parameters of these functions do not have any physical significance in terms of equivalence to process parameters such as heat or mass transfer coefficients, reaction kinetics, etc. This is the disadvantage of black box models compared to mechanistic models. However, if the aim is to merely represent faithfully some trends in process behaviour, then the black box modelling approach is just as effective. Moreover, the cost of modelling is orders of magnitude smaller than that associated with the development of mechanistic models.

As shown in Fig. 8A.1, black box models can be further classified into linear and nonlinear forms. In the linear category, transfer function and time series models predominate. With sampled data systems, this delineation is, in a sense, arbitrary. The only distinguishing factor is that in time-series models, variables are treated as random variables. In the absence of random effects, the transfer function and time-series models are equivalent. Given the relevant data, a variety of techniques may be used to identify the parameters of linear black box models [Eykhoff, 1974]. The most common techniques used, though, are least-squares based algorithms.

Under the nonlinear category, time-series feature again together with neural network based models. In nonlinear time-series, the nonlinear behaviour of the process is modelled by combinations of weighted cross-products and powers of the variables used in the representation. The parameters of the functions are still linear and thus facilitates identification using least squares based techniques. Neural networks are not new paradigms to nonlinear systems modelling. However, the increase in cheap computing power and certain powerful theoretical results have led to a resurgence in the use of neural networks in model building [Cybenko, 1989; Lippmann, 1987, Rummelhart and McCelland, 1986].

8A.5 QUALITATIVE MODELS

There are instances where the nature of the process may preclude mathematical description, e.g., when the process is operated at distinct operating regions or when physical limits exist. This results in discontinuities that are not amenable to mathematical descriptions. In this case, qualitative models can be formulated. The simplest form of a qualitative model is the 'rule-based' model that makes use of 'IF-THEN-ELSE' constructs to describe process behaviour. These rules are elicited from human experts. Alternatively, Genetic Algorithms and Rule Induction techniques can be applied to process data to generate these describing rules [South et al, 1993]. More sophisticated approaches make use of Qualitative Physics theory [Bobrow, 1984; Weld and deKleer, 1990] and its variants. These latter methods aim to rectify the disadvantages of purely rule based models by invoking some form of algebra so that the preciseness of mathematical modelling approaches could be achieved.

Of these, Qualitative Transfer Functions (QTFs) [Feray Beaumont et al, 1998] appear to be the most suitable for process monitoring and control applications. QTFs retain many of the qualities of quantitative transfer functions that describe the relationship between an input and an output variable, particularly the ability to embody temporal aspects of process behaviour. The technique was conceived for applications in the process control domain. Cast within an object framework, a model is built up of smaller sub-systems and connected together as in a directed graph. Each node in the graph represents a variable while the arcs that connect the nodes describe the influence or relationship between the nodes. Overall system behaviour is derived by traversing the graph, from input sources to output sinks.

Models derived based on the use of Fuzzy Set theory can also be classified as qualitative models. Proposed by Zadeh [1965, 1971], fuzzy set theory contains an algebra and a set of linguistics that facilitates descriptions of complex and ill-defined systems. Magnitudes of changes are quantised as 'negative medium', 'positive large' and so on. The model combines elements of the rule based and probabilistic approaches and sets of symbols with interpretations such as, '*If the increment of the input is positive large, the possibility of the increment on the output being negative small is 0.8*'. Fuzzy models are being used in everyday life without our being aware of their presence, e.g., washing machines, auto focus cameras, etc.

8A.6 STATISTICAL MODELS

Describing processes in statistical terms is another modelling technique. Time-series analysis which has a heavy statistical bias may be considered to fall into this model category. Nevertheless, due to its widespread and interchangeable use in the development of deterministic as well as stochastic digital control algorithms, the earlier classification is more appropriate. The statistical approach is made necessary by the uncertainties surrounding some process systems. This technique has roots in statistical data analysis, information theory, games theory and the theory of decision systems.

Probabilistic models are characterised by the probability density functions of the variables. The most common is the normal distribution which provides information about the likelihood of a variable taking on certain values. Multivariate probability density functions can also be formulated but interpretation becomes difficult when more than two variables are considered. Correlation models arise by quantifying the degree of similarity between two variables by monitoring their variations. This is again quite a commonly used technique, and is implicit when associations between variables are analysed using regression techniques.

System dynamics are not captured by statistical models. However, in modern control practice, they play an important role particularly in assisting in higher level decision making, process monitoring, data analysis and obviously, in Statistical Process Control.

8A.7 MODEL BASED (MODERN) AUTOMATIC CONTROL

Given a representative model of a process, 'What-If' investigations can be made via simulation, to answer operational questions such as safety related issues and to provide for operator training. However, this approach is not suitable for real-time automatic control. Within the context of automatic control, the inverse problem is considered, i.e., given the current states of the process, what actions should be taken to achieve desired specifications. Depending on the form of the plant model, different control strategies can be developed. The attraction of adopting a model based approach to controller development is illustrated in the block diagram shown in Figure 8A.2.

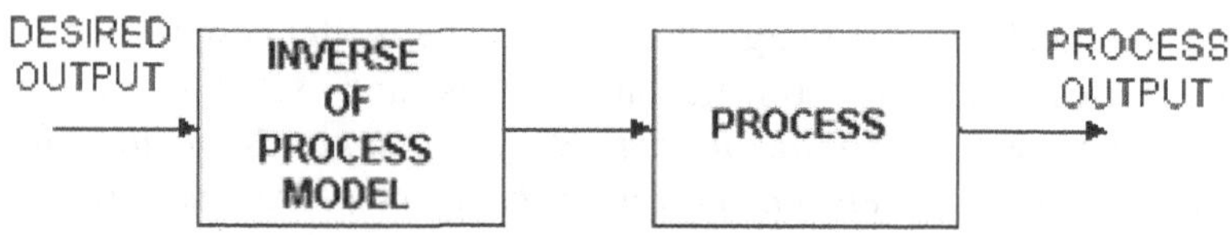

Figure 8A.2 Ideal Model Based Control.

By regarding the blocks to be mathematical operators, it can be seen that if an accurate model of the process is available, and if its inverse exists, then process dynamics can be cancelled by the inverse model. As a result, the output of the process will always be equal to the desired output. In other words, model based control design has the potential to provide perfect control. Hence, the first task in the implementation of modern control is to obtain a model of the process to be controlled. However, given that there are constraints on process operations; that all models will contain some degree of error and that all models may not be invertible, perfect control is very difficult to realise. These are the issues that modern control techniques aim to address, either directly or indirectly.

In the process industries, black box models are normally used for controller synthesis because the ill-defined nature of the processes makes mechanistic model development very costly. For process design purposes, precise characterisation is important. However, for the purposes of control strategy specification, controller design and control system analysis, models that can replicate the dynamic trends of the target processes are usually sufficient. Black box models have been found to be suitable in this respect and can be used to predict the results of taking certain actions.

Linear transfer functions and time-series descriptions are popular model forms used in control systems design. This is because of the wealth of knowledge that has been built up in linear systems theory. Increasingly, however, controllers are being designed using nonlinear time-series as well as neural network based models in recognition of the nonlinearities that pervade real world applications. The following sections briefly discuss the various algorithms that may arise from model based controller designs.

8A.7.1 PID CONTROL

The ubiquitous three-term Proportional + Integral + Derivative (PID) controller accounts for more than 80% of installed automatic feedback control devices in the process industries. In the past, these have been tuned using frequency response techniques or empirically derived rules-of-thumb. The modern approach is to determine the settings of the PID controller based upon a model of the process. The settings are chosen so that the controlled response adhere to user specifications. A typical criterion is that the controlled response should have a quarter decay ratio. Alternatively, it may be desired that the controlled response follow a defined trajectory or that the closed loop has certain stability properties [Warwick and Rees, 1988].

It can be easily shown that a Proportional + Integral controller is optimal for a first order linear process without time-delays. Similarly, the PID controller is optimal for a second order linear process without time-delays. In practice, process characteristics are nonlinear and can change with time. Thus the linear model used for initial controller design may not be applicable when process conditions change or when the process is operated at another region.

One solution is to have a series of stored controller settings, each pertinent to a specific operating zone. Once it is detected that the operating regime has changed, the appropriate settings are switched in. This strategy, called parameter- or gain-scheduled control, has found favour in applications to processes where the operating regions are changed according to a preset and constant pattern. In applications to continuous systems, however, the technique is not so effective.

A more elegant technique is to implement the controller within an adaptive framework. Here the parameters of a linear model are updated regularly to reflect current process characteristics [Warwick et al, 1987; Willis and Tham, 1989a]. These parameters are in turn used to calculate the settings of the controller as shown schematically in Fig. 8A.3.

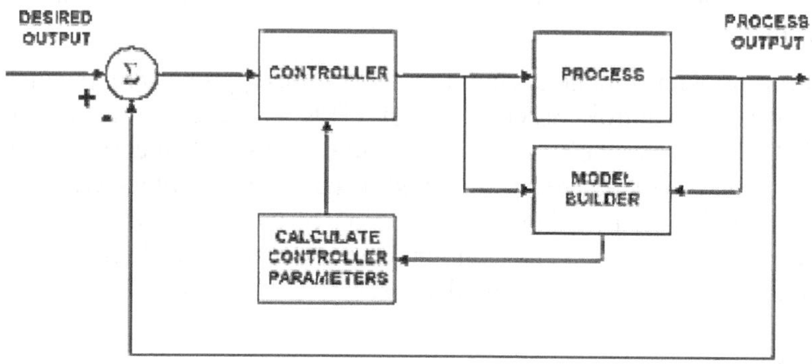

Figure 8A.3. Simplified Schematic of the Structure of Adaptive Controllers

The settings of the controller can be updated continuously according to changes in process characteristics. Such devices are therefore called auto-tuning/adaptive/self-tuning controllers. In some formulations, the controller settings are directly identified. A faster algorithm results because the model building stage has been avoided. Currently, many commercial auto-tuning PID controllers available from major control and instrumentation manufacturers. The simplest forms are those based upon the use of linear time-series models. Some PID controllers are also auto-tuned using pattern recognition methods [Bristol, 1977]. For example, the Foxboro EXACT controller changes its settings to maintain a user defined response pattern. A good review of auto-tuning PID controllers is given in Astrom and Hagglund [1988].

Theoretically, all model based controllers can be operated in an adaptive mode [Hang et al, 1993]. Nevertheless, there are instances when the adaptive mechanism may not be fast enough to capture changes in process characteristics due to system nonlinearities. Under such circumstances, the use of a nonlinear model may be more appropriate for PID controller design. Nonlinear time-series, and recently neural networks, have been used in this context. A nonlinear PID controller may also be automatically tuned using an appropriate strategy, by posing the problem as an optimisation problem. This may be necessary when the nonlinear dynamics of the plant are time-varying. Again, the strategy is to make use of controller settings most appropriate to the current characteristics of the controlled process. A self-tuning PID controller based on the use of a nonlinear neural net model has been reported by Montague and Willis (1993).

8A.7.2 PREDICTIVE CONSTRAINED CONTROL

PID type controllers do not perform well when applied to systems with significant time-delays. Perhaps the best known technique for controlling systems with large time-delays is the 'Smith predictor'. It overcomes the debilitating problems of delayed feedback by using predicted future states of the output for control. Currently, some commercial controllers have Smith predictors as programmable blocks. There are, however, many other model based control strategies have dead-time compensation properties. If there is no time-delay, these algorithms usually collapse to the PID form. Predictive controllers can also be embedded within an adaptive framework and a typical adaptive predictive control structure is shown in Fig. 8A.4.

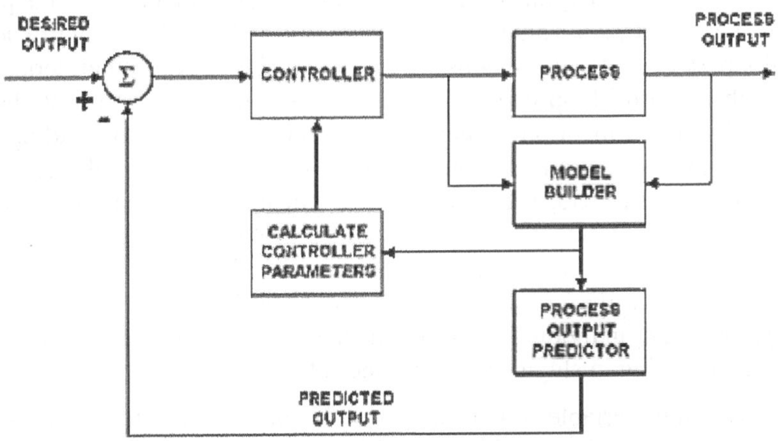

Figure 8A.4 Simplified Schematic of Adaptive Predictive Controllers

Classical Generalised Minimum Variance (GMV) controller is an example of this philosophy [Clarke and Gawthrop, 1975]. GMV control minimises the squared weighted difference between the desired value and the predicted output while penalising excessive control effort. The prediction horizon is the time-delay of the system, and this is a fixed parameter. GMV control, however, cannot effectively cope with variable time-delays and process constraints. This led to the development of long-range predictive controllers, e.g., the Generalised Predictive Controller (GPC) and Dynamic Matrix Control (DMC) [Clarke et al, 1987; Cutler and Ramaker, 1979; Wilkinson et al, 1990, 1991a,b]. The control problem is formulated in a manner similar to that adopted in the GMV approach. The differences are that the model is used to provide predictions of the output over a range of time-horizons into the future. Usually the range is between the smallest and largest expected delays. This alleviates the problem of varying time-delays and hence enhances robustness. Calculation of the control signal is essentially an optimisation problem. Here, economic objectives as well as process constraints can be included in the problem formulation. Examples of process constraints are the limits to liquid flows in fixed sized piping, allowable temperatures and pressures in process units, emissions to atmosphere, etc. Nowadays, the phrase 'predictive control' refers to the application of long-range predictive controllers. Again, predictive controllers may be designed using linear or nonlinear models.

8A.7.3 MULTIVARIABLE CONTROL

Thus far, we have only considered the case where there is one manipulated input and one controlled output; single-input single-output (SISO) systems. With most processes, there are many variables that have to be regulated. The chemical reactor is a typical example where level, temperature and pressure have to kept at design values, that is there are at least three control loops; a multi-loop system. If the actions of one controller affect other loops in the system, then control-loop interaction is said to exist. If each controller has been individually tuned to provide maximum performance, then depending on the severity of the interactions system instability may occur when all the loops are closed. SISO controllers, whether adaptive, linear or nonlinear strategies, may therefore not be applicable to such processes. Models used in the design of SISO controllers do not contain information about the effects of loop interactions. Thus, they cannot be expected to perform well. For a multiloop strategy to work, individual SISO controllers are usually detuned (made less sensitive), resulting in sluggish performances for some or all loops.

Ideally, multivariable controllers should be applied to systems where interactions occur. As opposed to multi-loop control, multivariable controllers take into account loop interactions and their de-stabilising effects. Fortunately, it is a relatively trivial task to modify model based controllers to accommodate multivariable systems. By regarding loop interactions as feed-forward disturbances, they can be easily included in the model description. This simple augmentation leads to multivariable linear decoupling controllers [Jones and Tham, 1987; Tham, 1985; Tham et al, 1991b; Vagi et al, 1991], as well as nonlinear neural network based multivariable control algorithms [Willis et al, 1991e]. Following SISO designs, multivariable controllers that can provide time-delay compensation and handle process constraints can also be developed with relative ease. By incorporating suitable numerical procedures to build the model on-line, adaptive multivariable control strategies result.

8A.7.4 ROBUST CONTROL AND THE INTERNAL MODEL PRINCIPLE

Using an on-line parameter estimation algorithm to identify the parameters of the model, the parameters of most linear model based controllers can be adjusted in line with changes in process characteristics. Although great strides have been made in resolving the implementation issues of adaptive systems, for one reason or other, many practitioners are still not confident about the long term integrity of the adaptive mechanism. This concern has led to another contemporary topic in modern control engineering; robust control.

Robust control involves, firstly, quantifying the uncertainties or errors in a 'nominal' process model, due to nonlinear or time-varying process behaviour for example. If this can be accomplished, we essentially have a description of the process under all possible operating conditions. The next stage involves the design of a controller that will maintain stability as well as achieve specified performance over this range of operating conditions. A controller with this property is said to be 'robust' [Morari and Zafiriou, 1989].

A sensitive controller is required to achieve performance objectives. Unfortunately, such a controller will also be sensitive to process uncertainties and hence suffer from stability problems. On the other hand, a controller that is insensitive to process uncertainties will have poorer performance characteristics in that controlled responses will be sluggish. The robust control problem is therefore formulated as a compromise between achieving performance and ensuring stability under assumed process uncertainties. Uncertainty descriptions are at best very conservative, whereupon performance objectives will have to be sacrificed. Moreover, the resulting optimisation problem is frequently not well posed. Thus, although robustness is a desirable property, and the theoretical developments and analysis tools are quite mature, application is hindered by the use of daunting mathematics and the lack of a suitable solution procedure.

Nevertheless, underpinning the design of robust controllers is the so called 'internal model' principle. It states that unless the control strategy contains, either explicitly or implicitly, a description of the controlled process, then either the performance or stability criterion, or both, will not be achieved. The corresponding 'internal model control' design procedure encapsulates this philosophy and provides for both perfect control and a mechanism to impart robust properties (see Fig. 8A.5).

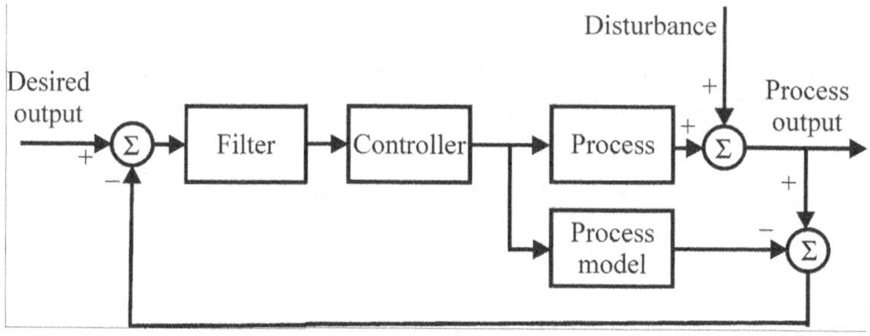

Figure 8A.5 Schematic of Internal Model Control Strategy

If the process model is invertible, then the controller is simply the inverse of the model. If the model is accurate and there is no disturbance, then perfect control is achieved if the filter is not present. This also implies that if we know the behaviour of the process exactly, then feedback is not necessary! The primary role of the low-pass filter is to attenuate uncertainties in the feedback, generated by the difference between process and model outputs and serves to moderate excessive control effort. The strategy and the concept that it embraces are clearly very powerful. Indeed, the internal model principle is the essence of model based control and all model based controllers can be designed within its framework.

8A.7.5 GLOBALLY LINEARISING CONTROL

As mentioned previously, there are cases when adaptive linear control schemes would not perform well when faced with a highly nonlinear process. This is because the adaptive mechanism may not be fast enough to track changes in process characteristics. Appropriately designed nonlinear controllers would therefore be expected to perform better. The use of neural network model based controllers has already been mentioned. Another emerging field is that of nonlinear controller designed based on mechanisitc models via the use of differential geometric concepts [Brockett, 1976; Kravaris and Kantor, 1990]. The aim of the design is similar to the use of Taylor series expansion to linearise the nonlinear model prior to application of linear model based controller designs. However, instead of providing local linearisation, contemporary nonlinear control strategies aim to provide 'global' linearisation over the space spanned by the states of the process; Globally Linearising Control (GLC). Global linearisation is achieved by a pre-compensator, designed such that the relationship between the inputs to the pre-compensator and the process output is linear. Linear control techniques can then be applied to the pseudo linear plant. A schematic of this strategy is shown in Fig. 8A.6.

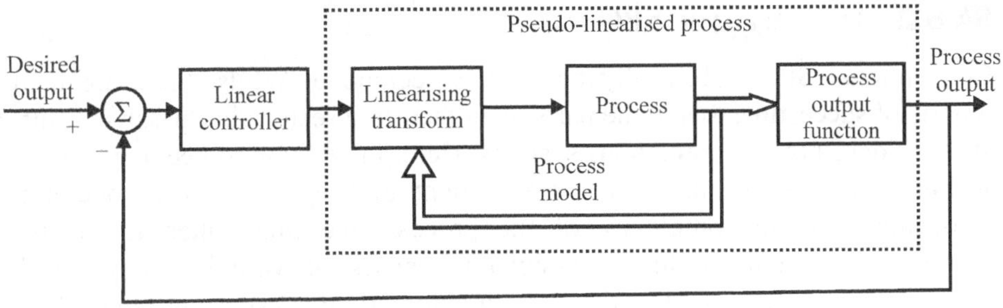

Figure 8A.6 Schematic of Globally Linearising Control

Globally linearising control is a relatively new development and much research is being still being carried out to investigate the applicability of the technique [e.g. McColm et al, 1994]. McLellan et al [1990] provide a review of nonlinear controller designs based upon mechanistic models. An interesting development that avoids the requirement of mechanistic models, is to use neural networks models instead. Neural network models are transformed into an equivalent state-space representation, and the GLC is designed based upon this state-space model [Peel et al, 1994].

8A.8 STATISTICAL PROCESS CONTROL

Statistical Process Control (SPC) is widely applied in the parts manufacturing industries. Although the technique has been practised at various levels for more than 30 years, it warrants mention. In response to current total quality initiatives SPC has only just recently begun to be implemented in the process industries. SPC makes use of statistical models and procedures that to improve product quality and process productivity at reduced costs [Wetherill and Brown, 1991]. The objective is to bring and keep processes in a state where any remaining variations are those inherent to the process.

8A.8.1 CONVENTIONAL SPC

SPC has been traditionally achieved by successive plotting and comparing a statistical measure of the variable with some user defined 'control' limits. If the plotted statistic exceeds these limits, the process is considered to be out of statistical control. Corrective action is then applied in the form of identification, elimination or compensation for the 'assignable' causes of variation. The most common charts used are the Shewhart, Exponential Moving Average (EWMA), range and Cumulative Sum (CuSum) charts.

8A.8.2 ALGORITHMIC SPC

Conventional SPC is basically an off-line technique. Whilst there are many reports of succesful cases in the parts manufacturing sector, this 'passive' control strategy does not suit continuous systems. Here, in addition to keeping products within specifications, there is a requirement to keep the process operating. Depending on the complexity of the process, the time taken to identify, eliminate and compensate for assignable causes of variation may not be acceptable. Nevertheless, the aim of both automatic process control and SPC is to increase plant profitability. Thus, it is reasonable to expect that the merger of these two apparently dichotomous methodologies could yield strategies that inherit the benefits associated with the parent approaches. This has been a subject of recent investigations [MacGregor, 1988; Tucker, 1989] where SPC is used to monitor the performances of automatic control loops. Such a strategy is sometimes called 'Algorithmic SPC' (ASPC), referring to the integrated use of algorithmic model based controllers and SPC techniques. Note, though, that the process is still being controlled by an automatic controller, that is the process is being controlled all the time.

8A.8.3 ACTIVE SPC

Another way to integrate the two control approaches is to provide on-line SPC. Statistical models are used not only to define control limits, but also to develop control laws that suggest the degree of manipulation to maintain the process under statistical control. Thus, in applications to continuous processes, the need for an algorithmic automatic controller is avoided, leading to a direct or 'active' SPC strategy [Efthimiadu and Tham 1990, 1991; Efthimiadu et al, 1993]. Indeed, the technique is designed specifically for continuous systems. In contrast to ASPC, manipulations are made only when necessary, as indicated by detecting violation of control limits. As a result, compared to automatic control and ASPC, savings in the use of raw materials and utilities can be achieved using active SPC.

8A.9 DEALING WITH DATA PROBLEMS

In the field of modern control engineering, much effort has been expended into the development and analysis of novel control strategies. A common assumption in these studies is that the required data is available. Unfortunately, this is often not the case in practice. However, the problem is gaining attention and a variety of solutions have been proposed to deal with various aspects of the difficulties associated with data.

8A.9.1 INFERENTIAL ESTIMATION

A major problem is the lack of on-line instrumentation to measure quantities that define product quality, e.g., stickiness of adhesives, smoothness of sheet material, melt flow index of polymers, flash points of fuels, etc. These are often provided by laboratory analyses resulting in infrequent feedback and substantial measurement delays, rendering automatic process control impossible. Inferential estimation is one method that has been designed to overcome this problem. The technique has also been called 'sensor-data fusion' and 'soft-sensing'.

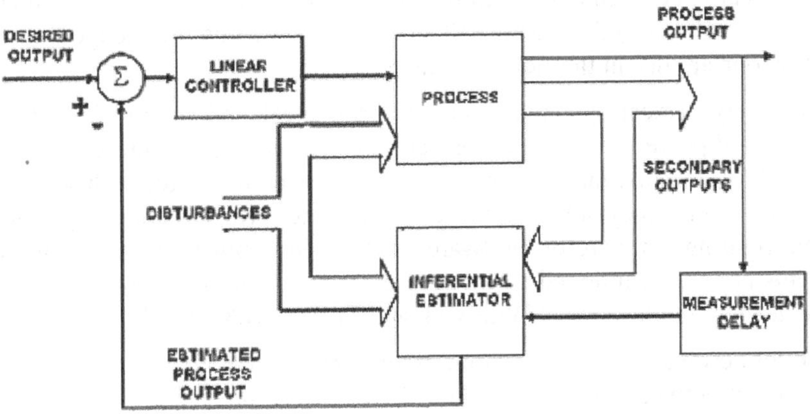

Figure 8A.7 Schematic of Feedback Control using Inferential Estimator.

Apart from the main quality variable, there are usually other variables such as temperatures, pressures, flows, etc., that are associated with a process. Changes in some of these variables are indicative of changes in product quality. Thus, by monitoring suitable secondary variables, it is often possible to 'infer' the state of the quality variable. Process operators and engineers do this on a daily basis in running process plants. However, the process may be complex and there could be many factors that affect product quality. As a result, the relationship between process conditions and product quality may not be straight forward, leading to inaccuracies in human judgement. Inferential Estimation alleviates this problem. The technique uses easily obtainable measurements of variables that are known to influence product quality, together with those of product quality when available, to generate estimates of product quality.

As with feedback control strategies, in applications to non-linear systems, the relationship between secondary variables and the primary output can be 'learnt' automatically. Thus, parameters that define the relationship are adjusted to match changes in process characteristics [Guilandoust et al, 1987, 1988; Lant et al, 1993; Montague et al, 1990, 1998; Tham, 1989, 1991a]. Alternatively, the inferential estimators may also be designed based upon the use of a neural network model. As shown in Fig. 8A.7, estimates of the quality which are generated at the measurement frequency of the secondary variables may then be used for process monitoring or control purposes.

8A.9.2 DATA CONDITIONING AND VALIDATION

Even if appropriate instruments exist, the data may not be of sufficient quality for desired goals to be achieved. Signals from plant are often corrupted by noise of varying magnitudes. All control methods are data driven. If appropriate measures are not taken to condition and validate the measured signals, then even the most sophisticated scheme will fail. In other words, the adage 'rubbish in, rubbish out' applies in the field of control.

In safety critical systems, such as the control and monitoring of nuclear reactors and power generators, steps are taken to ensure that the 'correct' signals are used for decision making. In these cases, it is common for both software and hardware redundancy schemes to be implemented. Redundancy is provided for by configuring software or hardware modules in duplicate or triplicate. Voting systems are then employed to validate output signals, retaining only those that are considered to be correct [Warwick and Tham, 1988, 1990].

In less critical applications, duplex or triplex redundancy configurations are not cost effective. Therefore, unless there is absolute need, the smoothing of noisy signals is accomplished via hardware or software filtering to attenuate noise in measured signals. However, a penalty is incurred if the signal is subject to spikes. To remove these spikes or rogue points, heavy filtering has to be applied whereupon significant time-lags may be introduced into the filtered signal. Time-lags in the filtered signal may however be reduced by employing 'logic' filters which combine conventional filter algorithms with SPC concepts to validate and condition process measurements. This integrated approach has been shown to be very effective [Parr and Tham, 1998; Tham and Parr, 1994]

8A.9.3 DATA ANALYSIS

Even if 'clean' data is available, there may be many variables associated with a particular process unit. The specification of an appropriate control strategy and controller design become complicated. Which variable should be manipulated to control another? What is the effect of this choice of manipulated-input controlled-output pairings? These are some of the important questions that have to be answered before a candidate configuration can be applied. Indeed, the results dictate the kind of models to employ for controller design and hence final controller types, and the overall control strategy that should be implemented. Inappropriate choice of input-output pairs exacerbate the problem of loop interactions. If interactions are significant, then a multivariable control design is necessary. If the input-output relations indicate nonlinear behaviour, then nonlinear controllers may have to be applied.

Many techniques can be used to tackle these issues. They range from simple graphical techniques (scatter plots, Box-plots), statistical multivariate analysis (e.g., Principal Component Analysis, Correlation Analysis, Cluster Analysis) to control theoretic, relative gain and singular value analyses. The latter are used to investigate control loop interactions and robustness of control strategies. However, there is currently no all embracing procedure for a systematic analysis of data right through to determining the suitability of the final control scheme.

8A.10 HIGHER LEVEL OPERATIONS

8A.10.1 PROCESS OPTIMISATION

The application of optimisation techniques is not restricted to the design of predictive constrained controllers. Process optimisation is a task in its own right. Unlike local controllers, which seek to maintain unit operating conditions at desired levels, the plant optimiser utilises a model of the plant to adjust operating conditions of the process so as to minimise raw material usage and maximise profits [Edgar and Himmelblau, 1989]. The outputs of the optimiser therefore set the targets for the local controllers, taking into consideration the operational limits of the plant. This effectively bridging the gap between the plant's true business objectives and its actual operations [Latour, 1979a, 1979b]. Figure 8A.8 shows a generic configuration of a process optimisation scheme.

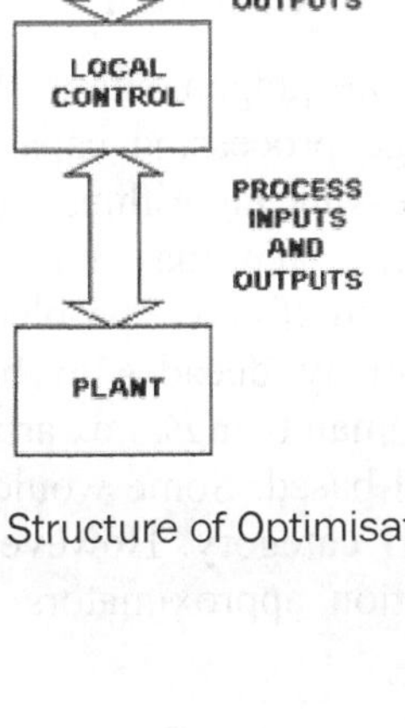

Figure 8A.8 Structure of Optimisation Scheme

Due to the complexity and the scale of this type of optimisation problem, the model used is normally a steady-state description to enable a tractable solution. As with control algorithms, adaptive on-line optimisation is also feasible [Bamberger and Isermann, 1978; Kambhampati et al, 1998; Willis and Tham, 1989b, 1990].

8A.11 PROCESS MONITORING, FAULT DETECTION, LOCATION AND DIAGNOSIS

Fault diagnosis has become an area of primary importance in modern process automation. It provides the pre-requisites for fault tolerance, reliability or security, which constitute fundamental design features in complex engineering systems. The system under consideration is monitored and the data is passed to fault detection algorithms or procedures. The basic task of a fault detection scheme is to register an alarm when an abnormal condition develops in the monitored system. Once a fault is detected, procedures may also be subsequently used to identify or diagnose the cause of the abnormality.

Fault detection and diagnosis techniques are again based upon the use of process models. In addition to the mathematical models used in controller design, statistical as well as qualitative models are increasingly being employed [Isermann, 1984; Patton et al, 1989]. Mathematical models are normally used to develop state-estimators or state-observers. Data from the monitored plant is input to these algorithms and the outputs compared with the corresponding plant outputs. If there are discrepancies, then it is an indication that at least one fault has occurred. The next task is to determine the locations of these faults. Again a representative model, not necessarily the one used in fault detection, is employed. In some instances, the location of the fault may be deduced by the type of fault. Here genetic algorithms and rule induction systems can be used to classify the fault.

8A.12 PROCESS SUPERVISION VIA ARTIFICIAL INTELLIGENCE TECHNIQUES

Human beings are able to make judgements in the face of subtle nuances and ambiguities. These knowledge processing capabilities cannot be matched by number crunching data processing algorithms, such as those described above. Although, the human decision system may not be precise, the result is often of sufficient accuracy for quick and effective problem solving. It has been the goal of computer scientists for many decades to build systems that mimic the decision making powers of human beings, i.e., artificial intelligent (AI) systems. AI techniques are also model based. Some would regard neural network based techniques to fall into the AI category. However, we tend to consider neural networks as numerical function approximators. Although AI techniques can make use of mathematical and statistical models, including neural networks, much of their utility is based upon the use of qualitative models.

Perhaps the most well known AI process supervisory schemes is based upon the use of expert systems [Efstathiou, 1986]. Expert systems are made up of three components. The rule or knowledge base holds information and logical rules for performing inference between facts. Next, there is the inference engine which controls the operation of the system and carries out the logical inference by processing the information in the knowledge base. The user interface makes up the final component, enabling communication between the user and the computer. Thus, an expert system is a collection of computer programs which operate upon the knowledge of experts in a particular application domain. Its purpose is to enable a novice to solve a problem with the benefits of the expert's knowledge.

When the inference engine and the user-interface are packaged as a single entity, this is known as an expert system shell. Software for procedures that can be combined together to form such a shell are known as expert system tools. The increasing availability of expert system shells and tools is a major reason for the proliferation of expert systems, where all that remains to be done is the compilation of the knowledge base. The extraction of rules that govern the operation of a process is called knowledge elicitation. This is performed via question and answer sessions between the extractor of knowledge, the so called knowledge engineer, and the provider of knowledge, the domain expert. There are also systems that are able to generate rules for expert systems when presented with data collected from a process. These are either based on rule induction techniques or genetic algorithms. However, the knowledge base could comprise any of the other qualitative models described previously and in any combination, including mathematical and statistical models.

When the system is presented with a collection of facts or a process scenario, the inference engine moves through the knowledge base in a 'forward' manner to come up with 'expert' advice or suggestions. However, unlike the implementation of 'IF-THEN-ELSE' constructs in conventional programming languages, expert systems have the ability to traverse the knowledge base in a backward direction. Backward chaining is invoked when the expert system is presented with a final result, and it is asked to provide a line of reasoning as to the events that led to the given result. Thus, another distinguishing factor of expert systems is that they are also able to provide explanations as to why a particular piece of advice or suggestion has been made. Expert systems have therefore found use in providing operator advice and as a process simulator for operator training [Kaemmerer and Christopherson, 1985].

Expert systems can be operated in two ways. The most common is the consultative mode where the expert system asks the user a series of questions.

Alternatively, the data required by the expert system is provided directly by interfacing to plant instruments. There is a growing number of expert system shells that can reason in real-time [Shaw, 1988]. Such Real-Time Knowledge Based Systems (RTKBS) have been used to tune controllers, supervise the performance of adaptive controllers, perform fault detection and diagnosis, perform alarm management and even provide direct on-line process control.

8A.13 ADVANCED CONTROL

The techniques described in the previous sections have been applied to a wide variety of systems. In the process industries, they have been applied to reactors, separation processes, power generation systems including boilers, HVAC and so on. Many of these are reported by academics, academics involved in industrial collaborative projects or by consultants. There are also many unreported cases of successful advanced control applications, primarily because of commercial confidentiality. An illustrative list of reported applications is given in Appendix A. Many of the applications reported in the literature describe the use of single techniques. However, our philosophy of advanced control is depicted in the following diagram.

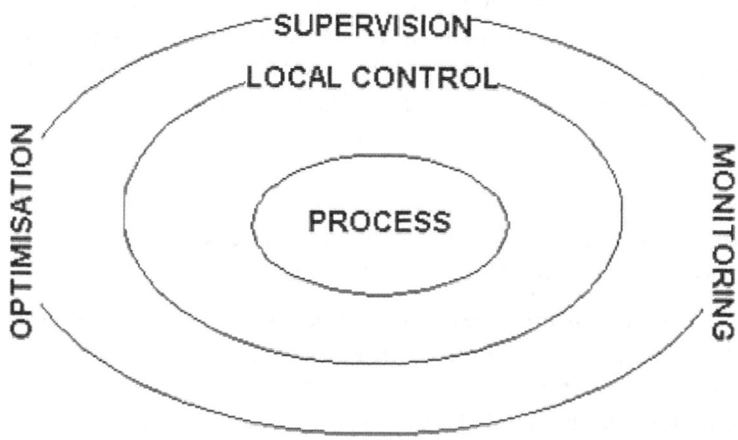

Figure 8A.9 Hierarchical Layers in Integrated Modern Control

Local control is implemented, using appropriate controllers, to keep the process operating at desired conditions. Here, the type of local controllers employed depends on the task at hand. Although it is easier to tune and maintain simple controllers, some processes do require control by more sophisticated algorithms. However, unless such sophisticated controllers are installed and maintained by well trained personnel, they can be prone to failure. Until the last decade, the higher level tasks of monitoring, optimisation, and supervision were mainly carried out by human beings. Due to the advent of modern technology, and advances in the field of AI, these can now be automated. In particular, the installation, operation and integrity of modern controllers can be supervised by higher level systems.

Advanced control is the implementation of this hierarchical information and control structure. The flow of information is bi-directional, from management layer to process level and vice versa. The task here is to be able to integrate the various components in an efficient and manageable fashion. This can be facilitated by ensuring that each component is designed as a modular, yet integrable element.

8A.14 CURRENT RESEARCH AND FUTURE TRENDS

In the process industries, the biggest challenge facing process engineers will be the reduction of variable costs whilst maintaining product quality. Advanced process control is the most effective technology available to realise this objective, especially on established plants. As systems become more complex, another important aspect is the reliability of the implemented systems. Here, the reliability of hardware and software are issues which have to be addressed. Allied to this is the requirement for suitably designed man-machine interfaces to enable efficient and reliable information transfer and to facilitate systems management.

With regard to the primary modules making up an advanced control project, neural networks, nonlinear systems theory, robust control, knowledge based systems are areas which appear to have captured the attention of both researchers and practitioners in the field of control engineering. This trend will continue well into the next decade. Areas that will receive particular attention will be techniques that will translate raw data into useful information; improved measurement methods including inferential estimation; multivariable non-linear predictive control and formal techniques for analysing the integrity of neural network based methodologies.

All information is of value, and should not be discarded just because they do not conform to a particular model building procedure. Thus, new modelling methods are also required. These should provide a framework where *a priori* knowledge of the process could be combined with the various existing modelling techniques, leading to so called 'grey-box' models. The resulting models should also be amenable for utilisation by the different modern controller designs, thus rendering controller synthesis independent of model types.

The process industries have an enormous base of manufacturing facilities which are still being run by unsophisticated or primitive control schemes. Competitive pressures will not allow any company in these industries to ignore the significant efficiencies possible through adopting modern process control technologies. A major obstacle to realising the full potential of modern control techniques is the lack of exposure to advances in the field. This can be overcome by the development of portable computer based training packages. The current proliferation in multi-media computing systems is the ideal impetus for the development of such learning aids.

REFERENCES

1. Anderson, J.S. (1998). 'Process control opportunities and benefits analysis'. Proc. Advanced Control for the Process Industries, Cambridge, 9-11th Sept.
2. Astrom, K.J. and Hagglund, T. (1988). Automatic Tuning of PID Controllers. ISA.
3. Bamberger, W. and Isermann, R. (1978). 'Adaptive on-line steady-state optimization of slow dynamic processes'. Automatica, Vol.14, pp.883-830.
4. Bobrow, D.G. (1984). 'Qualitative reasoning about physical systems: an introduction'. Artificial Intelligence, Vol.81, pp.1-5.
5. Bristol, E.H. (1977). 'Pattern recognition: an alternative to parameter identification in adaptive control', Automatica, Vol.13, pp.179-808.
6. Brockett, R.W. (1976). 'Nonlinear systems and differential geometry', Proc. IEEE, Vol.64, No.1, pp.61-71.
7. Brunet-Manquat, F., Willis, M.J. and Tham, M.T. (1993) 'Multi-Rate Adaptive Inferential Control', Submitted to Control'94
8. Clarke, D. and Gawthrop, P.J. (1975). 'Self-tuning controller'. Proc. IEE, Vol.188, pp.989-934.
9. Clarke, D., Mohtadi, C. and Tuffs, P.S. (1987). 'Generalised predictive control - part 1. The basic algorithm'. Automatica, Vol.83, No.8, pp.137-160.
10. Cutler, C.R. and Ramaker, B.L. (1979). 'Dynamic matrix control - a computer control algorithm', Proc. AIChE 86th Meeting, April.
11. Cybenko, G. (1989). 'Approximations by superpositions of a sigmoidal function', Math. Control Signal Systems, Vol.8, pp.303-314.
12. Di Massimo, C., Willis, M.J., Montague, G.A., Tham, M.T. and Morris, A.J. (1991). 'Bioprocess model building using artificial neural networks', Bioprocess Engineering, Vol.7, pp77-88.
13. DiMassimo, C., Lant, P.A., Saunders, A., Montague, G.A., Tham, M.T. and Morris, A.J. (1998). 'Bioprocess applications of model based estimation techniques'. J. Chem. Tech. Biotechnol., Vol.53, pp865-877.

14. DiMassimo, C., Montague, G.A., Willis, M.J., Morris, A.J. and Tham, M.T. (1998) 'Enhancing industrial bioprocess monitoring through artificial neural networks', ICCAFT 5 / IFAC BIO 8, Keystone, Colorado.

15. DiMassimo, C., Montague, G.A., Willis, M.J., Tham, M.T. and Morris, A.J. (1998). 'Towards improved penicillin fermentation via artificial neural networks'. Computers and Chemical Engineering, Vol.16, No.4, pp883-891.

16. Edgar, T.F. and Himmelblau, D. (1989). Optimization of Chemical Processes. McGraw-Hill.

17. Efstathiou, J. (1986). 'Introduction to expert systems'. Journal A, Vol.87, No.8, pp.57-61.

18. Efthimiadu, I. and Tham, M.T. (1990). 'Online statistical process control of continuous chemical processes', IEE Colloquium on 'Applied Statistical Process Control', Savoy Place, London, Nov.

19. Efthimiadu, I. and Tham, M.T. (1991). 'Online statistical process control of chemical processes'. Proc. IEE Conf., CONTROL'91, Heriot-Watt University, March 85-88.

20. Efthimiadu, I., Tham, M.T. and Willis, M.J. (1993). 'Engineering control and product quality assurance in the product industries'. Proc. Int. Conf. on Quality and It's Applications, Uni. Newcastle upon Tyne, Sept.

21. Eykhoff, P. (1974). System Identification. John Wiley.

22. Feray-Beaumont, S., Corea, R., Tham, M.T. and Morris, A.J. (1998). 'Process modelling for intelligent control'. J. Engineering Applications of Artificial Intelligence, Vol.5, pp.483-498.

23. French, I., Cox, C., Willis, M.J. and Montague, G.A. (1998) 'Intelligent Tuning of PI Controllers for Bioprocess Applications', AIRTC, Netherlands.

24. Glassey, J., Montague, G.A., Willis, M.J. and Morris, A.J. (1993) 'Considerations in Process Applications of Artifical Neural Networks', Colloquium on Neural Networks and Fuzzy Logic in Measurement and Control', Liverpool, March.

25. Guilandoust, M.T., Morris, A.J. and Tham, M.T. (1987). 'Adaptive inferential control'. Proc. IEE Pt.D, 134, 3, pp 171-179.

26. Guilandoust, M.T., Morris, A.J. and Tham, M.T. (1988). 'An adaptive estimation algorithm for inferential control'. I & EC Research, 87, pp 1658-1664.

27. Hang, C.C., Lee, T.H. and Ho, W.K. (1993). Adaptive Control. ISA.

28. Isermann, R. (1984). 'Process fault detection based on modeling and estimation methods - a survey'. Automatica, Vol.80, No.4, pp.387-304.

29. Jones, R.W. and Tham, M.T. (1987). 'Multivariable adaptive control: a survey of methods and applications'. In: Multivariable Control for Industrial Applications. (ed. J. O'Reilly), Peter Peregrinus, pp880-898.

30. Kaemmerer, W.F. and Christopherson, P.D. (1985). 'Using process models with expert systems to aid process control operators'. Proc. ACC'85, Boston, pp.898-897.

31. Kambhampati, C., Tham, M.T., Montague, G.A. and Morris, A.J. (1998). 'Optimsing control of fermentation processes'. Proc. IEE, Pt. D, Vol.139, No.1, pp 60-66.

32. Kravaris, C. and Kantor, J.C. (1990). 'Geometric methods for nonlinear process control - Parts 1 and 8'. Ind. Eng. Chem. Res., Vol.89, No.18, pp.8895-8383.

33. Kresta, J.V., MacGregor, J.F. and Marlin, T.E. ,(1991). 'Multivariate statistical monitoring of process operating performance'. Can. J. Chem. Eng., Vol.69, No. 35.

34. Lant, P.A., Tham, M.T. and Montague, G.A. (1993). 'On the applicability of adaptive bioprocess state estimators'. J. Bioeng. Biotech.

35. Lant, P.A., Willis, M.J., Montague, G.A., Tham, M.T. and Morris, A.J. (1990) 'A Comparison of Adaptive Estimation with Neural based techniques for Bioprocess Application', Preprints, ACC, San Deigo, pp 8173-8178

36. Latour, P.R. (1979a). 'On-line computer optimization 1: what it is and where to do it'. Hydrocarbon Processing, June, pp.73-88.

37. Latour, P.R. (1979b). 'On-line computer optimization 8: benefits and implementation'. Hydorcarbon Processing, July, pp.819-883.

38. Lippmann, R.P. (1987). 'An introduction to computing with neural nets'. IEEE ASSP Magazine, April.

39. MacGregor, J.F., (1988). 'Online statistical process control', Chem. Eng. Progress, Vol.81.

40. Manchanda, S., Willis, M.J., Tham, M.T., Montague, G.A., Morris, A.J. (1991) 'An appraisal of Nonlinear Control philosophies for application to a Biochemical process', ACC, Boston, USA.

41. McColm, E.J., Manchanda, S. and Tham, M.T. (1994). 'On the applicability of a nonlinear control strategy', Proc. IEE Conf. CONTROL'94, pp.1358-1357.

42. McLellan, P.J., Harris, T.J. and Bacon, D.W. (1990). 'Error trajectory descriptions of nonlinear controller designs'. Chemical Engineering Sci., Vol.45, No.10, pp.3017-3034.

43. Merry, M. (1983). 'APEX3: an expert system shell for fault diagnosis'. GEC J. of Research, Vol.1, No.1, pp.39-47.

44. Montague, G.A., DiMassimo, C., Willis, M.J., Tham, M.T. and Morris, A.J. (1990) 'Neural network modelling of a fermentation plant', Biotechnology Course, University of Edinburgh, March.

45. Montague, G.A., Morris, A.J. and Tham, M.T. (1998). 'Enhancing bioprocess operability with generic software sensors'. J. Biotechnology, 85, pp183-801.

46. Montague, G.A., Tham, M.T. and Lant, P.A. (1990). 'Estimating the immeasurable without mechanistic models'. Trends in Biotechnology, 8, 3, March, pp 88-83.

47. Montague, G.A., Tham, M.T., Willis, M.J. and Morris, A.J. (1998) 'Predictive control of distillation columns using dynamic neural networks', IFAC Conf. DYCORD+'98, Uni. of Maryland, College Park., U.S.A.

48. Montague, G.A., Willis, M.J., Di Massimo, C., Tham, M.T. and Morris, A.J. (1991) 'Application of neural networks in the control and modelling of industrial processes', IEE Neural Network Symposium, Poly. Central London., April.

49. Montague, G.A., Willis, M.J., Morris, A.J., (1991) 'Dynamic modelling of industrial processes with artificial neural networks.', 4th International Symposium. Neural Networks and Engineering Applications, University of Newcastle-upon-Tyne, Oct.

50. Montague, G.A., Morris, A.J. and Willis, M.J. (1998) 'Neural Networks (Methodologies for Process Modelling and Control)', AIRTC, Netherlands.

51. Montague, G.A., Morris, A.J. and Willis, M.J. (1998) 'Artificial Neural Networks: Methodologies and Applications in Process Control', Neural Networks for Control and Systems, Eds. Warwick, K., Irwin, G.W. and Hunt, K.J.

52. Montague, G.A., Willis, M.J., Tham, M.T. and Morris, A.J. (1991) 'Artificial Neural Network Based Predictive Control', IEE Control, 91', Edinburgh, March.

53. Montague, G.A., Willis, M.J., Tham, M.T., Morris, A.J., (1991) 'Artificial neural networks in estimation and control', IEE Coll. on Principles and Applications of Sensor Data Fusion, London, Feb.

54. Montague, G.A., Willis, M.J., Tham, M.T. and Morris, A.J. (1991) 'Artificial Neural Network Based Multivariable Predictive Control', 8nd Int. Conf. on Artificial Neural Networks, 18-80 Nov., Bournemouth.

55. Morari, M. and Zafiriou, E. (1989). Robust Process Control. Prentice-Hall.

56. Parr, A. and Tham, M.T. (1998). 'An on-line technique for the validation and reconstruction of data'. Proc. IChemE Research Event, Manchester, Jan.

57. Patton, R., Frank, P. and Clark, R. (1989). Fault Diagnosis is Dynamic Systems. Prentice Hall.

58. Peel, C., Willis, M.J. and Tham, M.T. (1998). 'A fast procedure for the training of neural networks'. J. Process Control, Vol.8, No.4.

59. Peel, C., Willis, M.J. and Tham, M.T. (1998). 'Application of a novel neural network training paradigm to industrial processes', Colloquium on the Application of Neural Networks to Modelling and Control', Liverpool, March.

60. Peel, C., Willis, M.J. and Tham, M.T. (1993). 'Enhancing Neural Network Training', Application of Neural Networks to Modelling and Control, Eds. Page, G.F., Gomm, J.B. and Williams, D.

61. Peel, C., Willis, M.J., Tham, M.T. and Manchanda, S. (1994). 'Globally linearising control using artificial neural networks', Proc. IEE Conf. CONTROL'94, pp.967-978.

CHAPTER **8B**

FUZZY LOGIC APPLICATION

A look at Fuzzy Logic as it relates to ABS (Anti-Lock Braking) within an automobile.

APPLICATION 1: FUZZY LOGIC AND ANTI-LOCK BRAKING SYSTEM

INTRODUCTION

In establishing a non-complex and practical application of the phenomenon of fuzzy logic it is important to consider a trifling situation in everyday life which is applicable to all, thus the event of anti-lock braking system was considered as a prime example of how we could demonstrate the implementation of fuzzy logic.

The result of this literature review will show the implementation of fuzzy logic in this relatively low-level control of some machinery as opposed to a high-level artificial intelligence application. This application is not uncommon as it has been seen where fuzzy logic has been implemented in relatively simple systems such as washing machines, traffic control, truck speed limiter, aircraft flight path and air conditioning to name a few. These implementations can be seen as an underachievement for the technology as fuzzy logic was originally developed to solve the complexities involved in the discipline of Artificial Intelligence. Mimicking the thought process of humans as well as doing language translations, were key areas that fuzzy logic was developed to be implemented.

FUZZY LOGIC

Fuzzy set theory, from whence fuzzy logic comes, was developed in 1965 by Lotfi Zadeh to combat the imprecision and uncertainties that exists in the everyday world. Its applications are geared towards solving non-mathematically distinct problems and allow statements to be answered with more than a YES or NO. Its derivative from traditional logic theorems, allows it to include all the properties of that system in addition to the new properties that were developed and hence, mapping functions, ordering and arithmetic operations all apply to fuzzy logic.

As opposed to its classical counterpart, fuzzy logic possesses exceedingly greater capabilities to capture uncertainties in their various forms, and as a result improves the gap between mathematical models and the associated physical reality. Fuzzy logic is capable of capturing the vagueness of linguistic terms in statements that are expressed in natural languages. Modeling human common sense reasoning, decision-making and other aspects of human cognition are enhanced with the use of fuzzy logic. These capabilities are essential in acquiring knowledge from human experts, representing and manipulating knowledge in expert systems in a human-like manner, and, generally in designing and building systems, which exhibit high levels of intelligence. Behavior, which is associated with perception rather than measurements, is an intriguing basis for fuzzy logic explorations.

Electronic control systems in the automotive industry are currently being pursued in the United States, and the reality of superior performance through the use of fuzzy logic based control rather than traditional control algorithms. Fuzzy logic strives to establish a value for linguistic expressions like *"fast"*, *"slow"* and *"long"* by finding an interval between 0% and 100% to accurately express the truthfulness of an expression. It also uses *"if, then"* rules to determine outcomes for particular input data. With this construct, it is possible to build rules such as:

> *"If the rear wheels are turning slowly and a short time ago the vehicle speed was high, then reduce rear brake pressure".*

Such rules provide themselves to the development of an ABS braking system based on fuzzy logic, and as such we shall proceed to describe this development as best as possible, given the predefined limit of this research document.

THE FUZZY ABS SYSTEM

ABS is implemented in automobiles to ensure optimal vehicle control and minimal stopping distances during hard or emergency braking. The number of cars equipped with ABS has been increasing continuously in the last few years. ABS is now accepted as an essential contribution to vehicle safety. The methods of control utilized by ABS are responsible for system performance.

Since ABS systems are nonlinear and dynamic in nature, they are a prime candidate for fuzzy logic control. For most driving surfaces, as vehicle braking force is applied to the wheel system, the longitudinal relationship of friction between vehicle and driving surface rapidly increases. Wheel slip under these conditions is largely considered to be the difference between vehicle velocity and a reduction of wheel velocity during the application of braking force. Brakes work because friction acts against slip. The more slip, given enough friction, the more braking force is brought to bear on the vehicles momentum. Unfortunately, slip can and will work against itself during cornering or on wet or icy surfaces where the coefficient of surface friction varies. If braking force continues to be applied beyond the driving surface useful coefficient of friction, the brake effectively begins to operate in a non-friction environment. Increasing brake force in a decreasing frictional environment often results in full wheel lockup. It has been both mathematically and empirically proven a sliding wheel produces less friction than a moving wheel.

HOW DOES IT WORK?

Conventional ABS control algorithms must account for non-linearity in brake torque due to temperature variation and dynamics of brake fluid viscosity. Also, external disturbances such as changes in frictional coefficient and road surface must be accounted for, not to mention the influences of tire wear and system components aging. These influential factors increase system complexity, in turn, affecting mathematical models used to describe systems. As the model becomes increasingly complex, equations required to control ABS also become increasingly complicated. Due to the highly dynamic nature of ABS many assumptions and initial conditions are used to make control achievable. Once control is achieved the system is physically implemented and tested. The system is then modified to achieve the desired control status.

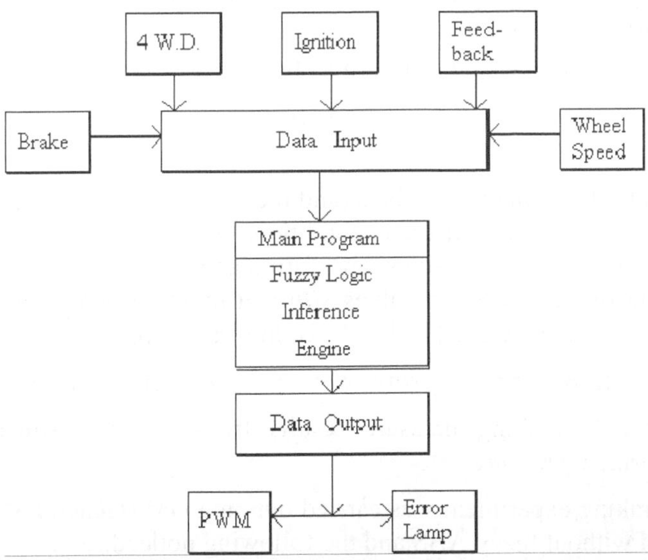

Fig. 8B.1

Inputs: The inputs to a particular Fuzzy ABS are represented in the Fig. 8B.1 and consist of:

1. ***The Brake:*** This block represents the brake pedal deflection/assertion. This information is acquired in a digital or analog format.
2. ***The 4 W.D:*** This indicates if the vehicle is in the 4-wheel-drive mode. However the vehicle may not be limited strictly to 4 W.D, as there exists vehicles that are based on Front-Wheel, Rear-Wheel or All-Wheel drive platforms (which is in-fact different from 4 W.D).
3. ***The Ignition:*** This input registers if the ignition key is in place, and if the engine is running or not.
4. ***Feed-back:*** This block represents the set of inputs concerning the state of the ABS system.
5. ***Wheel speed:*** In a typical application this will represent a set of 4 input signals that convey the information concerning the speed of each wheel. This information is used to derive all necessary information for the control algorithm.

THE FUZZY CONTROLLER

One such Fuzzy-Controller uses input values, which were converted to fuzzy variables *slip* and *dvwheel* by the fuzzification process. Both variables use seven linguistic values, the slip variable is described by the terms:

Slip = {*zero, very small, too small, smaller than optimum, optimum, too large, very large*},

The acceleration *dvwheel* is described by:

dvwheel = {*negative large, negative medium, negative small, negative few, zero, positive small, positive large*}.

As a result of two fuzzy variables (and the laws of probability), each of them having seven labels, 49 different conditions are possible. The rule base is complete, meaning, all 49 rules are formulated and all 49 conditions are allowed. The two fuzzy input values *slip* and *dvwheel* can be mapped to the fuzzy output value *pressure*. The labels for this value are:

pressure = {*positive fast, positive slow, zero, negative slow, negative fast*}

The optimal breaking pressure results from the defuzzification of the linguistic variable *pressure*.

A full-braking experiment was carried out on a GM (General Motors) motor car with and without fuzzy ABS and the following noticed:

Without fuzzy ABS the braking pressure reaches a very high level and wheel lockup occurred within a short space of time. Resulting in an unstable behavior, the vehicle cannot be steered anymore and the stopping distance increases.

With fuzzy ABS controller activated, steering was not only retained during the whole braking maneuver, but the slowing down length was considerably shortened as well.

CONCLUSION

On the basis of better control under probable life threatening and dangerous conditions, which may occur during the braking process within an automobile, it is safe to conclude on the basis of our research/literary review, that with the advent of the incorporation of fuzzy logic within the electronic braking system of an automobile, safer braking conditions have been achieved. The ABS system has been for a number of years, used by the automotive industry, and with further research, development and implementation of more refined fuzzy logic concepts, the driving world will forever more, be safer during their daily commute, knowing that such systems are in place to help avoid skidding and other hazardous mechanisms which they may face.

GLOSSARY

- Fuzzification – the process by which input values are translated into linguistic concepts, which are represented by fuzzy sets.
- Defuzzification – the process by which the linguistic representation of data by fuzzy sets, are reconverted to the original input data.

REFERENCES

1. Eichfeld, H., Leindl, R. and Künemund, T.: *The SAE 81C99A Fuzzy Logic Coprocessor*, Embedded Intelligence ´96, ed. J. Wiesböck, Design & Elektronik, Sindelfingen 1996, pp. 347-353.
2. Elting, David and Robert Kowalczyk (et al). "Fuzzy Anti-Lock Brake System Solution", http://www.intel.com/design/mcs96/designex/2351.htm.
3. Klein, Ralph (et al). "Antilock-Braking System and Vehicle Speed Estimation using Fuzzy Logic", http://www.fuzzyTech.com/.
4. Symans, Micheal D. & Kelly, Stephen W. "Fuzzy Control of Bridge Systems…" 1999. Robert V. Demicco, George J. Klir. "Fuzzy Logic in Geology". 2003.

APPLICATION 2: WEB-BASED FUZZY NEURAL NETWORKS FOR STOCK PREDICTION

Abstract: A web-based stock prediction system is developed based on a fuzzy neural network by using the past stock data to discover fuzzy rules and make future predictions. The learning algorithm is implemented. Input data to each network are the moving averages of the weekly stock data, which are obtained from http://moneycentral.msn.com. The output simulation data are also the average values of the weekly stock data. After the input data are collected from the website for the specific term using web search techniques, the system is trained and then is able to make future predictions. To implement this stock prediction System, JSP (Java Server Page), JDK1.3, JSP server and IE server 5.0 are used.

INTRODUCTION

The financial market different from a lot of physical systems like we know the weather is, that the financial market is a sort of complex feedback mechanism. What people expect prices to be, affects the prices they observe and then the prices they observe then affects how they are going to form their expectations about what the prices will be in the next period. The market is basically an uncertain beast or an uncertain institution, it is an institution where people trade risk, swap risk, and that's why it is there. And so if it were possible to predict it there would be no risk. In individuals, I think there cannot be any publicly available system to predict a financial market. On the other hand, neural networks have been found useful in stock price prediction [1-2]. Both feedforward and recurrent neural networks have been investigated and good results have been obtained. That means the prediction software would be very useful to assist individuals in reaching a final decision. In this paper, assuming that it is possible to predict markets, a prediction system is developed using fuzzy neural networks with a learning algorithm to predict the future stock values. The system consists of several neural networks modules. These models are all used to learn the relationships between different technical and economical indices and the decision to buy or sell stocks. The inputs to the networks are technical and economic indices. The output of the system is the decision to buy and sell. There are several neural network methods for stock prediction, such as Time Series method, Recurrent neural network and Feed-forward neural network method, etc. [2]. When compared to these techniques, Fuzzy neural network is a very useful and effective method to process, which is explained in the later sections.

The learning algorithm is used to train the networks. Before learning starts, tolerances are defined for the output units. During learning, the weights are updated only when the output errors exceed the tolerances. The learning data for which the output errors do not exceed the tolerances are eliminated from the training data sets. The input data to each network are the moving averages of the weekly averaged data which are obtained directly by using a Java program from the website. The output simulation data is also the average values of the weekly stock data.

MODEL ANALYSIS

Time series forecasting analyzes past data and projects estimates of future data values. Basically, this method attempts to model a nonlinear function by a recurrence relation derived from past values. The recurrence relation can then be used to predict new values in the time series, which hopefully will be good approximations of the actual values. There are two basic types of time series forecasting: univariate and multivariate. Univariate models, like Box-Jenkins, contain only one variable in the recurrence equation. The equations used in the model contain past values of moving averages and prices. Box-Jenkins is good for short-term forecasting but requires a lot of data, and it is a complicated process to determine the appropriate model equations and parameters. Multivariate models are univariate models expanded to "discover casual factors that affect the behavior of the data." [3-4]. As the name suggests, these models contain more than one variable in their equations. Regression analysis is a multivariate model, which has been frequently compared with neural networks. Overall, time series forecasting provides reasonable accuracy over short periods of time, but the accuracy of time series forecasting diminishes sharply as the length of prediction increases. Many other computer-based techniques have been employed to forecast the stock market. They range from charting programs to sophisticated expert systems. Fuzzy logic has also been used. Expert systems process knowledge sequentially and formulate it into rules. They can be used to formulate trading rules based on technical indicators. In this capacity, expert systems can be used in conjunction with neural networks to predict the market. In such a combined system, the neural network can perform its prediction, while the expert system could validate the prediction based on its well-known trading rules. The advantage of expert systems is that they can explain how they derive their results. With neural networks, it is difficult to analyze the importance of input data and how the network derived its results. However, neural networks are faster because they execute in parallel and are more fault tolerant.

The major problem with applying expert systems to the stock market is the difficultly in formulating knowledge of the markets because we ourselves do not completely understand them. Neural fuzzy networks have an advantage over expert systems because they can extract rules without having them explicitly formalized. In a highly chaotic and only partially understood environment, such as the stock market, this is an important factor. It is hard to extract information from experts and formalize it in a way usable by expert systems. Expert systems are only good within their domain of knowledge and do not work well when there is missing or incomplete information. Neural networks handle dynamic data better and can generalize and make "educated guesses." Thus, neural networks are more suited to the stock market environment than expert systems. In the wide variety of different models presented so far, each model has its own benefits and shortcomings. The best way is that these methods work best when employed together. The major benefit of using a fuzzy neural network then is for the network to learn how to use these methods in combination effectively, and hopefully learn how the market behaves as a factor of our collective consciousness.

FUZZY NEURAL NETWORKS

ARCHITECTURE

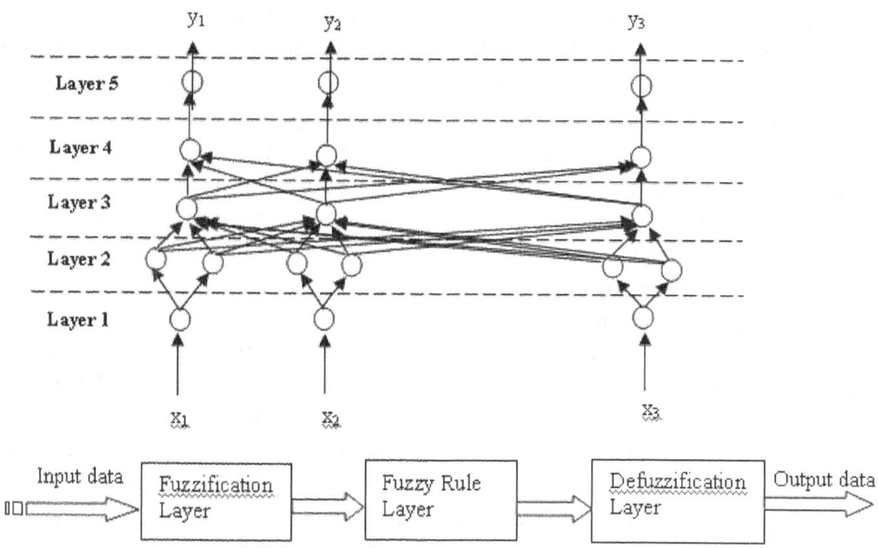

Fig. 8B.2 Architecture of the fuzzy neural network

According to the mechanism of fuzzy logic control system, the fuzzy neural network usually has 5 functional layers: (1) Layer 1 is the input layer. (2) Layer 2 is the fuzzification layer; (3) Layer 3 is the fuzzy reasoning layer which may consist of AND layer and OR layer; (4) Layer 4 is the defuzzification layer; (5) Layer 5 is the output layer. The architecture of a fuzzy neural network is described in Fig. 8B.2. Usually, the fuzzy neural network maps crisp inputs

x_i (I = 1,2,...,n) to crisp output y_i (j = 1,2,...,m). A fuzzy neural network is constructed layer by layer according to linguistic variables, fuzzy IF-THEN rules, the fuzzy reasoning method and the defuzzification scheme of a fuzzy reasoning method and the defuzzification scheme of a fuzzy logic control system.

Each neuron in the fuzzification layer represents an input membership function of the antecedent of a fuzzy rule. One common method to implement this layer is to express membership functions as discrete points. Thus for a fuzzy rule "IF X1 is A1 and X2 is A2 ... THEN Y is B", A's characterize the possibility distribution of the antecedent clause "X is A". Each of the hidden nodes is defined as a fuzzy reference point in the input space. The function of the defuzzification layer is for rule evaluation. Each neuron in this layer represents a consequent proposition "THEN Y is B" and its membership function can be implemented by combining one or two sigmoid functions and linear functions.

LEARNING ALGORITHM

An n-input-1-output fuzzy neural network has m fuzzy IF-THEN rules which are described by

$$\text{IF } x_1 \text{ is } A_1^k \text{ and } \ldots \text{ and } x_n \text{ is } A_n^k \text{ THEN } y \text{ is } B_n^k,$$

where x_i and y are input and output fuzzy linguistic variables, respectively. Fuzzy linguistic values A_i^k and B^k are defined by fuzzy membership functions as follows,

$$\mu_{A_k^i}(x_i) = \exp[-(\frac{x_i - a_i^k}{\sigma_i^k})^2] \qquad \text{.....(8B.1)}$$

$$\mu_{B^k}(y) = \exp[-(\frac{y - b^k}{\eta^k})^2] \qquad \text{.....(8B.2)}$$

the n-input-1-output fuzzy neural network with simple fuzzy reasoning is defined below:

$$f(x_1,\ldots,x_n) = \frac{\sum_{k=1}^{m} b^k [\prod_{i=1}^{n} \mu_{A_i^k}(x_i)]}{\sum_{k=1}^{m} [\prod_{i=1}^{n} \mu_{A_i^k}(x_i)]} \qquad \text{.....(8B.3)}$$

Given n-dimensional input data vectors x^p (*i.e.*, $x^p = (x_1^p, x_2^p, \ldots\ldots, x_n^p)$) and one-dimensional output data vector y^p for p=1,2,...,N, (i.e., N training data sets).

The energy function for p is defined by

$$E^{p} = \frac{1}{2}[f(x_1^p,...,x_n^p) - y^p]^2 \qquad .....(8B.4)$$

For simplicity, let E and f^p denote E^p and $f(x_1^p,...,x_n^p)$, respectively. After training the centers of output membership functions ($\frac{\partial E^P}{\partial b^k}$), the widths of output membership functions($\frac{\partial E^P}{\partial \delta^k}$), the centers of input membership functions($\frac{\partial E^P}{\partial a^k}$) and the centers of input membership functions($\frac{\partial E^P}{\partial \sigma^k}$), then we obtain the training algorithm [5 -7]:

$$b^k(t+1) = b^k(t) - \theta \left.\frac{\partial E^P}{\partial b^k}\right|_t \qquad .....(8B.5)$$

$$\sigma^k(t+1) = \sigma^k(t) - \theta \left.\frac{\partial E^P}{\partial \sigma^k}\right|_t \qquad .....(8B.6)$$

$$a^k(t+1) = a^k(t) - \theta \left.\frac{\partial E^P}{\partial a^k}\right|_t \qquad .....(8B.7)$$

$$\eta^k(t+1) = \eta^k(t) - \theta \left.\frac{\partial E^P}{\partial \eta^k}\right|_t \qquad .....(8B.8)$$

where, η is the learning rate and t = 0,1,2,...

The main steps using the learning algorithm as follows:

Step 1: Present an input data sample, compute the corresponding output;

Step 2: Compute the error between the output(s) and the actual target(s);

Step 3: The connection weights and membership functions are adjusted;

Step 4: At a fixed number of epochs, delete useless rule and membership function nodes, and add in new ones;

Step 5: IF Error > Tolerance THEN go to Step 1 ELSE stop.

When the error level drops to below the user-specified tolerance, the final interconnection weights reflect the changes in the initial fuzzy rules and membership functions. If the resulting weight of a rule is close to zero, the rule can be safely removed from the rule base, since it is insignificant compared to others. Also, the shape and position of the membership functions in the Fuzzification and Defuzzification Layers can be fine tuned by adjusting the parameters of the neurons in these layers, during the training process.

SYSTEM IMPLEMENTATION

INPUT DATA AND SOFTWARE PARAMETERS

The system can predict future for any stock or market index. For example to predict Microsoft stock values for some days ahead, historical data is required. Data for neural networks is probably the most important aspect for training. Without user intervention a particular stock data is obtained from the Internet directly as user's requirement, in this process, user can determine which stock and how long of price information he wants to get. Preprocessing operation is needed as preparatory step for next stage. As an example of preprocessing, the downloadable data is daily price, it is need to be grouped by week, and the average of each week are calculated for the next step. A HTML parser that coded in java is used to retrieve data. This program parses the whole file line by line to get the needed information and insert them into the database. All of training data are arranged for 2-input-1-output system as this format: (d1, d2 ---d3), (d2, d3 ---d4), where, for the first vector, d1 and d2 are inputs and d3 is an output, and in the second vector, d2 and d3 are inputs and d4 is an output.

There are some parameters, which we have to use during the training period. They are error, threshold, tolerance, etc. The accurate results depend on these parameters. To achieve our target rapidly, some parameters are preprocessed. For example, in the fuzzy member function 1 and 2, there are two important parameters: center and width of the fuzzy set. Their initial values have critical impact on the performance of system. The random value generated by machine is not optimal. In this system, we propose a simple way to optimize the initial value: sort all of data obtained from internet, and then divide data into five groups, get the average value of each group, these values are used in member function as the initial value of fuzzy set center.

The prediction algorithm takes all these parameters as input. This algorithm is called when we click predict future from the system after entering the stock symbol. This algorithm makes the neural network learn. The algorithm returns the future values as output which are eventually stored in results table for each stock. We will keep track of these results until we click on the average error for all simulations for this particular stock. Java program displays all the results of predictions for any particular stock at a time on the web page and clears the contents of the table.

OVERVIEW OF IMPLEMENTATION

A full run of the program implementation will be described, going through all the main features of the program:

1. Download historical data from the Internet.
2. A program is written through which user can get the data as his/her requirement. Arrange all of data into the format that used for the next step, and insert data into database.
3. Algorithm, which trains the neural networks using the mean square error as stop criterion for learning, while never exceeding the maximum number of cycles which can take testing data from the initial date to the user entered date, and predicts the future stock closing values.
4. A program is written, which compares the predicted values with real values.

One of the most important factors to construct a neural network depends on what the network will learn. A neural network must be trained on some input data. The two major problems in implementing the training are: defining the set of input to be used (the learning environment) and deciding on an algorithm.

SIMULATIONS

Using the developed system to predict the future stock values with Fuzzy neural Networks we can do some analysis to know the performance of the Back-Propagation Algorithm.

By using the past historical data, if we predict stock values for future 5 weeks from Back-Propagation algorithm we are now able to compare the predicted values with the real values. Table 8B.1 and Figure 8B.3 show the prediction and real values of the weekly average of the stock of I2 Technologies, Inc. The input past historical data is from 2001.1 to 2001.5. Table 8B.2 and Figure 8B.4 show the prediction and real values of the weekly average of the stock of Cypress Semiconductor Corporation. The input past historical data is from 2000.6 to 2001.1. The simulation results are very close to the real values.

When we changed the number of the input past historical data, the more input data we have the better training and get more close results. This means that, more the available data for predicting financial markets, the greater the chances of an accurate forecast. For the same test case if we decrease the maximum training error parameter from 0.0001 to 0.000015, we are getting more close results. The resulting future stock values are closer than the future stock values with high training error.

Simulations are done on several other stocks, such as Microsoft, Oracle and IBM, etc. From the simulation results it is conclusive that, the average error for simulations using lot of data is small compared to the average error using less data and the more data for training the neural network, the better prediction it gives.

Table 8B.1 The weekly average values of the stock of I2 Technologies, Inc(from 2001.5)

	1st week	2nd week	3rd week	4th week	5th week
Web-based	20.763	19.851	19.386	19.2	19.116
Real valued	21.874	20.77	17.568	17.24	17.712

Table 8B.2 The weekly average values of the stock of Cypress Semiconductor Corporation(from 2001.1)

	1st week	2nd week	3rd week	4th week	5th week
Web-based	25.014	24.227	23.658	23.326	23.139
Real valued	25.008	23.338	23.872	20.575	19.358

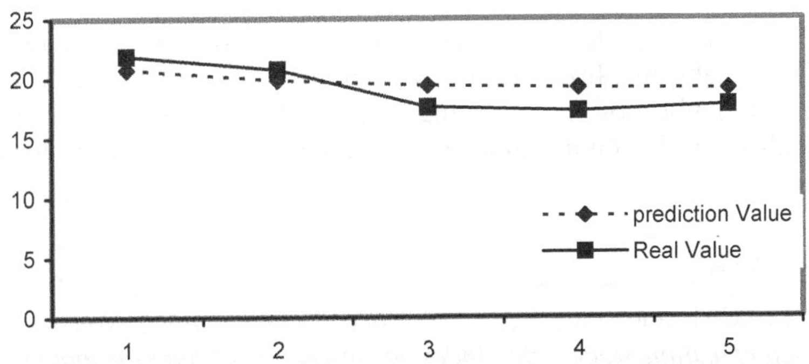

Fig. 8B.3 The simulation values and real values of the stock of I2 Technologies, Inc(from 2001.5)

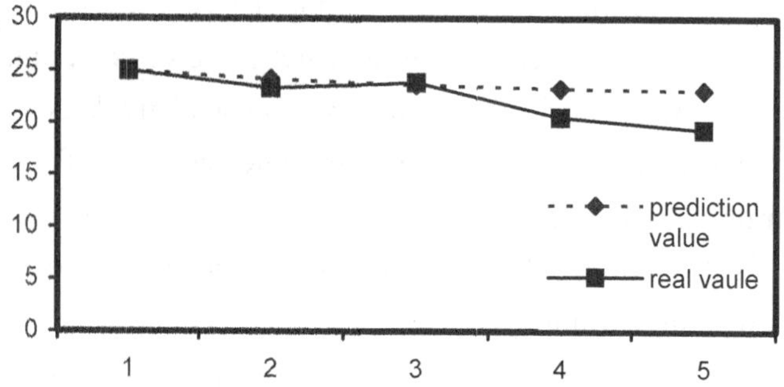

Fig. 8B.3 The simulation values and real values of the stock of Cypress Semiconductor Corporation (from 2001.1)

CONCLUSION

After completing several simulations for predicting several stocks based on the past historical data using fuzzy neural network with the Back-Propagation learning algorithm, it is conclusive that the average error for simulations using lots of data is smaller than that using less amount of data. That is, the more data for training the neural network, the better prediction it gives. If the training error is low, predicted stock values are close to the real stock values.

One possibility for future work is to update the system, so that it can read the past stock data automatically from the web. In this way the system will become Internet ready for predicting any stock market and is ready at any time. Another possibility for future work is to update the system, which can allow to trade the stock, which means users can manage to buy and sell the stock after seeing the prediction values from this system. In this way the system can eventually become Internet ready to be used anywhere in the world at any time. It is also possible to make the fuzzy neural Web-based stock prediction system as a commercial application by updating such that it gives more user-friendly functionality and by giving more valuable information to the users on palm machine.

REFERENCES

1. Lee,C.H and Park,K.C., "*Prediction of monthly transition of the composition stock price index using recurrent back-propagation*", Int.. Conf. On Artificial Neral Networks, Brighton,UK, pp.1629-1632, 1992.

2. D.T.Pham and X.Liu, "*Neural Networks for Identification, Prediction and Control*", Springer, 1999.

3. J. Ellman, "*Finding structure in time*", Cognitive Science, pages 179-211, 1990.

4. J. Robert and Van Eyden, "*The Application of Neural Networks in the Forecasting of Share Prices*," Finance and Technology Publishing, 1996.

5. Witold Pedrycz, Abraham Kandel and Yan-Qing Zhang, "Neurofuzzy Systems," In Fuzzy Systems: Modeling and Control, pages 312–363, Kluwer Academic Publishers, 1999.

6. Y.-Q. Zhang, M. D. Fraser, R. A. Gagliano and A. Kandel, "*Granular Neural Networks for Numerical-Linguistic Data Fusion and Knowledge Discovery*," Special Issue on Neural Networks for Data Mining and Knowledge Discovery, *IEEE Transactions on Neural Networks*, Vol. 11, No. 3, pp.658-667, May, 2000.

7. Y.-Q. Zhang and A. Kandel, "Compensatory Genetic Fuzzy Neural Networks and Their Applications," Series in Machine Perception Artificial Intelligence, Volume 30, World Scientific, 1998.

APPLICATION 3

A NEURO-FUZZY SYSTEM FOR SPEECH RECOGNITION

Abstract: Neural networks are excellent classifiers, but performance is dependent on the quality and quantity of training samples presented to the network. In cases where training data is sparse or not fully representative of the range of values possible, incorporation of fuzzy techniques improves performance. That is, introducing fuzzy techniques allow the classification of imprecise data. The neuro-fuzzy system presented in this study is a neural network that processes fuzzy numbers. By incorporating this attribute, the system acquires the capacity to correctly classify imprecise input.

Experimental results show that the neuro-fuzzy system's performance is vastly improved over a standard neural network for speaker-independent speech recognition. Speaker independent speech recognition is a particularly difficult classification problem, due to differences in voice frequency (amongst speakers) and variations in pronunciation. The network developed in this study has an improvement of 45% over the original multi-layer perceptron used in a previous study.

INTRODUCTION

Neural network performance is directly related to the size and quality of training samples [1]. When the number of training pairs is small, or perhaps not representative of the possibility space, standard neural network results are predictably poor. Incorporation of fuzzy techniques can improve performance in these cases [2]. Even though standard neural networks are excellent classifiers, introducing fuzzy techniques allow them to classify imprecise data. The neuro-fuzzy system is presented is a standard neural network modified to process fuzzy numbers [3, 4].

Processing fuzzy numbers can be accomplished in a variety of ways. One of the most elegant methods, because of its simplicity, is by using interval mathematics. The neuro-fuzzy system, then, is a standard feed-forward neural net that has been modified to deal with fuzzy numbers via interval mathematics. The modifications that are entailed are basically generalizations in the learning rule and neuronal functions in order to accommodate the interval mathematics.

THE NEURO-FUZZY SYSTEM

Data Representation and Processing

This neuro-fuzzy system is based on the neural network described by Ishibushi et. al. [5]. It was originally presented as a neural network that learned from fuzzy If-Then rules. This network configuration can be used in several ways, the key to which is taking α-cuts of the fuzzy number in question and utilizing interval mathematics. Specifically, α-cuts of the fuzzy numbers are represented by interval vectors [4]. That is, an α-cut of fuzzy input vector is represented by the interval vector $\mathbf{X}_p = (X_{p1}, X_{p2}, ..., X_{pn})^T$ where

$$X_{pi} = [x_{pi}^L, x_{pi}^U] \quad .....(8B.9)$$

indicate the lower and upper limits of the interval. Thus, the input and output vectors are interval vectors, and the neuronal operations are modified to deal with the interval numbers. Specifically, summation of weighted inputs is carried out as

$$\text{Net}_{pj}^L = \sum_{\substack{i \\ w_{ji} \geq 0}} w_{ji} o_{pi}^L + \sum_{\substack{i \\ w_{ji} < 0}} w_{ji} o_{pi}^U + \theta_j \quad .....(8B.10)$$

and

$$\text{Net}_{pj}^U = \sum_{\substack{i \\ w_{ji} \geq 0}} w_{ji} o_{pi}^U + \sum_{\substack{i \\ w_{ji} < 0}} w_{ji} o_{pi}^L + \theta_j \,. \quad .....(8B.11)$$

These calculations are consistent with the interval multiplication operation described in Alefeld [3]. The output equation can then be expressed as

$$\begin{aligned} o_{pj} &= [o_{pj}^{L}, o_{pj}^{U}] \\ &= [f(Net_{pj}^{L}),\ f(Net_{pj}^{U})] \end{aligned} \qquad \text{.....(8B.12)}$$

Derivation of the Learning Rule

The learning rule for this network must then be modified accordingly. Succinctly stated, the Generalized Delta (backpropagation with momentum) rule to update any weight is

$$\Delta w_{ji}(t+1) = \eta\left(-\frac{\partial E_p}{\partial w_{ji}}\right) + \alpha \Delta w_{ji}(t)\,. \qquad \text{.....(8B.13)}$$

The error is computed as the difference between the target output, $\mathbf{t}_p$, and the actual output, $\mathbf{o}_p$:

$$E_p = \max\{\frac{1}{2}(t_{pj} - o_{pj})^2\,, o_{pj} \in \mathbf{o}_p\}, \qquad \text{.....(8B.14)}$$

where

$$(t_{pj} - o_{pj}) = \begin{cases} (t_{pj} - o_{pj}^{L}), & \text{if } t_p = 1, \text{ and} \\ (t_{pj} - o_{pj}^{U}), & \text{if } t_p = 0. \end{cases} \qquad \text{.....(8B.15)}$$

For units in the output layer, calculation of $\partial E_p / \partial w_{ji}$ is straightforward, and can be thought of as four cases based on the value of target output and weight. Note that in the four equations the value of j in the subscript is fixed (to the output neuron that had the maximum error). In the first case,

the $t_p = 1$, and $w_{ji} \geq 0$:

$$\begin{aligned} \frac{\partial E_p}{\partial w_{ji}} &= \frac{\partial}{\partial w_{ji}}\left[\frac{(t_{pj} - o_{pj}^{L})^2}{2}\right] \\ &= \frac{\partial}{\partial o_{pj}^{L}}\left[\frac{(t_{pj} - o_{pj}^{L})^2}{2}\right]\frac{\partial o_{pj}^{L}}{\partial Net_{pj}^{L}} \cdot \frac{\partial Net_{pj}^{L}}{\partial w_{ji}} \\ &= -(t_{pj} - o_{pj}^{L}) \cdot o_{pj}^{L} \cdot (1 - o_{pj}^{L}) \cdot o_{pi}^{L} \\ &= -\delta_{pj}^{L} \cdot o_{pi}^{L}\,. \end{aligned} \qquad \text{.....(8B.16)}$$

The third line in the derivation assumes that the neuronal activation function, f (Net), is the binary sigmoid function, and thus substitutes the values accordingly. The other cases are evaluated with $t_p = 1$, and $w_{ji} < 0$, $t_p = 0$, and $w_{ji} \geq 0$, and $t_p = 0$, and $w_{ji} < 0$. This results in four output layer and eight hidden layer equations. The four output layer equations are summarized as follows:

$$\frac{\partial E_{pj}}{\partial w_{ji}} = \begin{cases} -\delta^L_{pj} \cdot o^L_{pi}, & \text{for} \quad t_p = 1, \text{ and } \quad w_{ji} \geq 0, \\ -\delta^L_{pj} \cdot o^U_{pi}, & \text{for} \quad t_p = 1, \text{ and } \quad w_{ji} < 0 \\ -\delta^U_{pj} \cdot o^U_{pi}, & \text{for} \quad t_p = 0 \text{ and } \quad w_{ji} \geq 0. \\ -\delta^U_{pj} \cdot o^L_{pi}, & \text{for} \quad t_p = 0 \text{ and } \quad w_{ji} < 0. \end{cases} \qquad \text{.....(8B.17)}$$

where

$$\delta^L_{pj} = (t_{pj} - o^L_{pj}) \cdot o^L_{pj} \cdot (1 - o^L_{pj}) \text{ and } \delta^U_{pj} = (t_{pj} - o^U_{pj}) \cdot o^U_{pj} \cdot (1 - o^U_{pj}).$$

The calculation of the partial derivative $\partial E_p / \partial w_{ji}$ for the hidden layers is based on back-propagating the error as measured at the output layer. The following discussion assumes one hidden layer, although subsequent terms could be derived in the same way for other hidden layers. Since this derivation involves a path of two neurons (output and hidden layer), there are eight cases. For the first case, $t_p = 1$, $w_{kj} \geq 0$ and $w_{ji} \geq 0$, where w_{kj} is the weight on the path from hidden layer neuron j to output layer neuron k, k fixed:

$$\frac{\partial E_p}{\partial w_{ji}} = \frac{\partial}{\partial w_{ji}} \left[\frac{(t_{pk} - o^L_{pk})^2}{2} \right]$$

$$= \frac{\partial}{\partial o^L_{pk}} \left[\frac{(t_{pk} - o^L_{pk})^2}{2} \right] \frac{\partial o^L_{pk}}{\partial \mathrm{Net}^L_{pk}} \cdot \frac{\partial \mathrm{Net}^L_{pk}}{\partial o_{pj}} \cdot \frac{\partial o^L_{pj}}{\partial \mathrm{Net}^L_{pj}} \cdot \frac{\partial \mathrm{Net}^L_{pj}}{\partial w_{ji}}$$

$$= -\delta^L_{pk} \cdot w_{kj} \cdot o^L_{pj} \cdot (1 - o^L_{pj}) \cdot o^L_{pi}. \qquad \text{.....(8B.18)}$$

The second case is a variation on the first, in which $t_p = 1$, $w_{kj} \geq 0$ and $w_{ji} < 0$, so the partial derivative becomes:

$$\frac{\partial E_p}{\partial w_{ji}} = -\delta^L_{pk} \cdot w_{kj} \cdot o^L_{pj} \cdot (1 - o^L_{pj}) \cdot o^U_{pi}. \qquad \text{.....(8B.19)}$$

The third case is characterized by $t_p = 1$, $w_{kj} < 0$ and $w_{ji} \geq 0$:

$$\frac{\partial E_p}{\partial w_{ji}} = -\delta^L_{pk} \cdot w_{kj} \cdot o^U_{pj} \cdot (1 - o^U_{pj}) \cdot o^U_{pi} \ . \qquad \text{.....(8B.20)}$$

$t_p = 1$, $w_{kj} < 0$ and $w_{ji} < 0$ for the fourth case:

$$\frac{\partial E_p}{\partial w_{ji}} = -\delta^L_{pk} \cdot w_{kj} \cdot o^U_{pj} \cdot (1 - o^U_{pj}) \cdot o^L_{pi} \ . \qquad \text{.....(8B.21)}$$

The fifth through eighth cases deal with a target value of 0. For the fifth case, $t_p = 0$, $w_{kj} \geq 0$ and $w_{ji} \geq 0$:

$$\frac{\partial E_p}{\partial w_{ji}} = \frac{\partial}{\partial w_{ji}}\left[\frac{(t_{pk} - o^U_{pk})^2}{2}\right]$$

$$= \frac{\partial}{\partial o^U_{pk}}\left[\frac{(t_{pk} - o^U_{pk})^2}{2}\right] \frac{\partial o^U_{pk}}{\partial Net^U_{pk}} \cdot \frac{\partial Net^U_{pk}}{\partial o^U_{pj}}$$

$$\cdot \frac{\partial o^U_{pj}}{\partial Net^U_{pj}} \cdot \frac{\partial Net^U_{pj}}{\partial w_{ji}}$$

$$= -\delta^U_{pk} \cdot w_{kj} \cdot o^U_{pj} \cdot (1 - o^U_{pj}) \cdot o^U_{pi} \ . \qquad \text{.....(8B.22)}$$

$t_p = 0$, $w_{kj} \geq 0$ and $w_{ji} < 0$ for the sixth case:

$$\frac{\partial E_p}{\partial w_{ji}} = -\delta^U_{pk} \cdot w_{kj} \cdot o^U_{pj} \cdot (1 - o^U_{pj}) \cdot o^L_{pi} \ . \qquad \text{.....(8B.23)}$$

In the seventh case, $t_p = 0$, $w_{kj} \geq 0$ and $w_{ji} < 0$:

$$\frac{\partial E_p}{\partial w_{ji}} = -\delta^U_{pk} \cdot w_{kj} \cdot o^L_{pj} \cdot (1 - o^L_{pj}) \cdot o^L_{pi} \ . \qquad \text{.....(8B.24)}$$

The eighth and final case has $t_p = 0$, $w_{kj} < 0$ and $w_{ji} < 0$:

$$\frac{\partial E_p}{\partial w_{ji}} = -\delta^U_{pk} \cdot w_{kj} \cdot o^L_{pj} \cdot (1 - o^L_{pj}) \cdot o^U_{pi} \ . \qquad \text{.....(8B.25)}$$

Equations (8B.17) through (8B.25) are used to code the training function in the simulation. Simulation results are presented in the next section.

IMPLEMENTATION AND EXPERIMENTAL RESULTS

The implementation of the neuro-fuzzy network was carried out as a simulation. Using the equations derived in the previous section, the simulation was coded in the C programming language and tested with several classic data sets. These results proved favorable, with comparable or improved performance over a standard neural network [6].

Vowel Recognition Data

The application problem that will serve as a testbench for this study is *Vowel*, one of a collection of data sets used as a neural network benchmarks [7]. It is used for speaker-independent speech recognition of the eleven vowel sounds from multiple speakers. The *Vowel* data set used in this study was originally collected by Deterding [8], who recorded examples of the eleven steady state vowels of English spoken by fifteen speakers for a "non-connectionist" [7] speaker normalization study. Four male and four female speakers were used to create the training data, and the other four male and three female speakers were used to create the testing data. The actual composition of the data set consists of 10 inputs, which are obtained by sampling, filtering and carrying out linear predictive analysis.

Specifically, the speech signals were low pass filtered at 4.7 kHz and digitized to 12 bits with a 10 kHz sampling rate. Twelfth order linear predictive analysis was carried out on six 512 sample Hamming windowed segments from the steady part of the vowel. The reflection coefficients were used to calculate 10 log area parameters, giving a 10 dimensional input space [7]. Each speaker thus yielded six frames of speech from eleven vowels. This results in 528 frames from the eight speakers used for the training set, and 462 frames from the seven speakers used to create the testing set.

528 samples is relatively small training set (in standard neural network applications), the values are diverse and thus *Vowel* is an excellent testbench for the neuro-fuzzy system.

Performance Results

As stated before, speaker-independent voice recognition is extremely difficult, even when confined to the eleven vowel sounds that *Vowel* consists of. Robinson carried out a study comparing performance of feed-forward networks with different structures using this data, which serve as a baseline for comparisons. The best recognition rate for a multi-layer perceptron reported in that study is 51% [9]. In spite of its difficulty, other studies have also used the Vowel data set. For various types of systems, the reported recognition rates (best case) are 51% to 59% [9, 10, 11]. The recognition rate obtained with the neuro-fuzzy system is 89%. Results of the simulation are summarized in Table 8B.3.

A recognition rate of 89% surpassed expectations, especially with a data set as diverse as the speaker-independent speech (vowel recognition) problem.

Table 8B.3 Best Recognition Rates for Standard Neural Networks and the Neuro-Fuzzy System

Type of Network	Number of Hidden Neurons	Best Recognition Rate
Std. Neural Network	11	59.3
Std. Neural Network	22	58.6
Std. Neural Network	88	51.1
Neuro-Fuzzy System	11	88.7

CONCLUSIONS

Fuzzy theory has been used successfully in many applications [12]. This study shows that it can be used to improve neural network performance. There are many advantages of fuzziness, one of which is the ability to handle imprecise data. Neural networks are known to be excellent classifiers, but their performance can be hampered by the size and quality of the training set. By combining some fuzzy techniques and neural networks, a more efficient network results, one which is extremely effective for a class of problems. This class is characterized by problems with its training set. Either the set is too small, or not representative of the possibility space, or very diverse. Thus, standard neural network solution is poor, because the training sequence does not converge upon an effective set of weights. With the incorporation of fuzzy techniques, the training converges and the results are vastly improved.

One example of the problem class which benefits from neuro-fuzzy techniques is that of speaker-independent speech recognition. The *Vowel* data set was used in simulation experiments, as reported in the previous section. This data set is well known and has been used in many studies, and has yielded poor results. This is true to such a degree that it caused one researcher to claim that "poor results seem to be inherent to the data" [10]. This difficulty with poor performance adds more credence to the effectiveness of the fuzzy techniques used in the neuro-fuzzy system.

In summation, the neuro-fuzzy system outperforms standard neural networks. Specifically, the simulations presented show that the neuro-fuzzy system in this study outperforms the original multi-layer perceptron study by 45% on the difficult task of vowel recognition.

REFERENCES

1. Kosko, Bart, *Neural Networks and Fuzzy Systems: A Dynamical Approach to Machine Intelligence*, Prentice Hall, Englewood Cliffs, NJ, 1992.

2. Nava, P. and J. Taylor, "The Optimization of Neural Network Performance through Incorporation of Fuzzy Theory," Proceedings of The Eleventh International Conference on Systems Engineering, pp. 897-901.

3. Alefeld, G. and J. Herzberger, *Introduction to Interval Computation*, Academic Press, New York, NY, 1983.

4. Bojadziev, G. and M. Bojadziev, *Fuzzy Sets, Fuzzy Logic, Applications*, World Scientific, New Jersey, 1995.

5. Ishibuchi, H., R. Fujioka, and H. Tanaka, "Neural Networks That Learn from Fuzzy If- Then Rules," Vol.1, No.2, pp. 85-97, 1993.

6. Nava, P., *New Paradigms for Fuzzy Neural Networks,* Ph.D. Dissertation, New Mexico State University, Las Cruces, New Mexico, 1995.

7. Carnegie Mellon University Neural Network Benchmark Collection, Pittsburgh, PA, 1994.

8. Deterding, D.H. *Speaker Normalization for Automatic Speech Recognition*, Ph.D. Dissertation, Cambridge, England, 1989.

9. Robinson, A.J., *Dynamic error Propagation Networks*, Ph.D. Dissertation, Cambridge University, Cambridge, England, 1989.

10. Graña, M., A. D'Anjou, F. Albizuri, J. Lozano, P. Laranaga, M. Hernandez,

11. F. Torrealdea, A. de la Hera, and I. Garcia, "Experiments of fast learning with High Order Boltzmann Machines," Informatica y Automatica, 1995.

12. Nava, P. and J. Taylor, "Voice Recognition with a Fuzzy Neural Network," Proceedings of The Fifth IEEE International Conference on Fuzzy Systems, pp. 2049-2052.

13. Zadeh, L.A., *Fuzzy Sets and Applications: Selected Papers by L. A. Zadeh*, John Wiley & Sons, New York, 1987.

APPENDIX - A

FUZZY LOGIC: AN INTRODUCTION TO FUZZY LOGIC

FUZZY LOGIC - AN INTRODUCTION - PART 1

Introduction

This is the first in a series of six articles intended to share information and experience in the realm of fuzzy logic (FL) and its application. This article will introduce FL. Through the course of this article series, a simple implementation will be explained in detail. Each article will include additional outside resource references for interested readers.

Where Did Fuzzy Logic Come From?

The concept of Fuzzy Logic (FL) was conceived by Lotfi Zadeh, a professor at the University of California at Berkley, and presented not as a control methodology, but as a way of processing data by allowing partial set membership rather than crisp set membership or non-membership. This approach to set theory was not applied to control systems until the 70's due to insufficient small-computer capability prior to that time. Professor Zadeh reasoned that people do not require precise, numerical information input, and yet they are capable of highly adaptive control. If feedback controllers could be programmed to accept noisy, imprecise input, they would be much more effective and perhaps easier to implement. Unfortunately, U.S. manufacturers have not been so quick to embrace this technology while the Europeans and Japanese have been aggressively building real products around it.

What is Fuzzy Logic?

In this context, FL is a problem-solving control system methodology that lends itself to implementation in systems ranging from simple, small, embedded micro-controllers to large, networked, multi-channel PC or workstation-based data acquisition and control systems. It can be implemented in hardware, software, or a combination of both. FL provides a simple way to arrive at a definite conclusion based upon vague, ambiguous, imprecise, noisy, or missing input information. FL's approach to control problems mimics how a person would make decisions, only much faster.

How is Fl Different from Conventional Control Methods?

FL incorporates a simple, rule-based IF X AND Y THEN Z approach to a solving control problem rather than attempting to model a system mathematically. The FL model is empirically-based, relying on an operator's experience rather than their technical understanding of the system. For example, rather than dealing with temperature control in terms such as "SP =500F", "T <1000F", or "21 °C < TEMP < 22 °C", terms like "IF (process is too cool) AND (process is getting colder) THEN (add heat to the process)" or "IF (process is too hot) AND (process is heating rapidly) THEN (cool the process quickly)" are used. These terms are imprecise and yet very descriptive of what must actually happen. Consider what you do in the shower if the temperature is too cold: you will make the water comfortable very quickly with little trouble. FL is capable of mimicking this type of behavior but at very high rate.

How Does Fl Work?

FL requires some numerical parameters in order to operate such as what is considered significant error and significant rate-of-change-of-error, but exact values of these numbers are usually not critical unless very responsive performance is required in which case empirical tuning would determine them. For example, a simple temperature control system could use a single temperature feedback sensor whose data is subtracted from the command signal to compute "error" and then time-differentiated to yield the error slope or rate-of-change-of-error, hereafter called "error-dot". Error might have units of degs F and a small error considered to be 2F while a large error is 5F. The "error-dot" might then have units of degs/min with a small error-dot being 5F/min and a large one being 15F/min. These values don't have to be symmetrical and can be "tweaked" once the system is operating in order to optimize performance. Generally, FL is so forgiving that the system will probably work the first time without any tweaking.

SUMMARY

FL was conceived as a better method for sorting and handling data but has proven to be a excellent choice for many control system applications since it mimics human control logic. It can be built into anything from small, hand-held products to large computerized process control systems. It uses an imprecise but very descriptive language to deal with input data more like a human operator. It is very robust and forgiving of operator and data input and often works when first implemented with little or no tuning.

REFERENCES

1. "Europe Gets into Fuzzy Logic" (Electronics Engineering Times, Nov. 11, 1991).
2. "Fuzzy Sets and Applications: Selected Papers by L.A. Zadeh", ed. R.R. Yager et al. (John Wiley, New York, 1987).
3. "U.S. Loses Focus on Fuzzy Logic" (Machine Design, June 21, 1990).
4. "Why the Japanese are Going in for this 'Fuzzy Logic'" by Emily T. Smith (Business Week, Feb. 20, 1993, pp. 39).

FUZZY LOGIC - AN INTRODUCTION - PART 2

Introduction

This is the second in a series of six articles intended to share information and experience in the realm of fuzzy logic (FL) and its application. This article will continue the introduction with a more detailed look at how one might use FL. A simple implementation will be explained in detail beginning in the next article. Accompanying outside references are included for interested readers.

In the last article, FL was introduced and the thrust of this article series presented. The origin of FL was shared and an introduction to some of the basic concepts of FL was presented. We will now look a little deeper.

Why Use Fl?

FL offers several unique features that make it a particularly good choice for many control problems.

1. It is inherently robust since it does not require precise, noise-free inputs and can be programmed to fail safely if a feedback sensor quits or is destroyed. The output control is a smooth control function despite a wide range of input variations.

2. Since the FL controller processes user-defined rules governing the target control system, it can be modified and tweaked easily to improve or drastically alter system performance. New sensors can easily be incorporated into the system simply by generating appropriate governing rules.
3. FL is not limited to a few feedback inputs and one or two control outputs, nor is it necessary to measure or compute rate-of-change parameters in order for it to be implemented. Any sensor data that provides some indication of a system's actions and reactions is sufficient. This allows the sensors to be inexpensive and imprecise thus keeping the overall system cost and complexity low.
4. Because of the rule-based operation, any reasonable number of inputs can be processed (1-8 or more) and numerous outputs (1-4 or more) generated, although defining the rulebase quickly becomes complex if too many inputs and outputs are chosen for a single implementation since rules defining their interrelations must also be defined. It would be better to break the control system into smaller chunks and use several smaller FL controllers distributed on the system, each with more limited responsibilities.
5. FL can control nonlinear systems that would be difficult or impossible to model mathematically. This opens doors for control systems that would normally be deemed unfeasible for automation.

How is Fl used?

1. Define the control objectives and criteria: What am I trying to control? What do I have to do to control the system? What kind of response do I need? What are the possible (probable) system failure modes?
2. Determine the input and output relationships and choose a minimum number of variables for input to the FL engine (typically error and rate-of-change-of-error).
3. Using the rule-based structure of FL, break the control problem down into a series of IF X AND Y THEN Z rules that define the desired system output response for given system input conditions. The number and complexity of rules depends on the number of input parameters that are to be processed and the number fuzzy variables associated with each parameter. If possible, use at least one variable and its time derivative. Although it is possible to use a single, instantaneous error parameter without knowing its rate of change, this cripples the system's ability to minimize overshoot for a step inputs.
4. Create FL membership functions that define the meaning (values) of Input/Output terms used in the rules.

5. Create the necessary pre- and post-processing FL routines if implementing in S/W, otherwise program the rules into the FL H/W engine.
6. Test the system, evaluate the results, tune the rules and membership functions, and retest until satisfactory results are obtained.

Linguistic Variables

In 1973, Professor Lotfi Zadeh proposed the concept of linguistic or "fuzzy" variables. Think of them as linguistic objects or words, rather than numbers. The sensor input is a noun, e.g. "temperature", "displacement", "velocity", "flow", "pressure", etc. Since error is just the difference, it can be thought of the same way. The fuzzy variables themselves are adjectives that modify the variable (e.g., "large positive" error, "small positive" error ,"zero" error, "small negative" error, and "large negative" error). As a minimum, one could simply have "positive", "zero", and "negative" variables for each of the parameters. Additional ranges such as "very large" and "very small" could also be added to extend the responsiveness to exceptional or very nonlinear conditions, but aren't necessary in a basic system.

SUMMARY

FL does not require precise inputs, is inherently robust, and can process any reasonable number of inputs but system complexity increases rapidly with more inputs and outputs. Distributed processors would probably be easier to implement. Simple, plain-language IF X AND Y THEN Z rules are used to describe the desired system response in terms of linguistic variables rather than mathematical formulas. The number of these is dependent on the number of inputs, outputs, and the designer's control response goals.

REFERENCES

1. "Clear Thinking on Fuzzy Logic" by L.A. Bernardinis (Machine Design, April 23, 1993).
2. "Fuzzy Fundamentals" by E. Cox (IEEE Spectrum, October 1992, pp. 58-61).
3. "Fuzzy Logic in Control Systems" by C.C. Lee (IEEE Trans. on Systems, Man, and Cybernetics, SMC, Vol. 20, No. 2, 1990, pp. 404-35).
4. "Fuzzy Sets" by Ivars Peterson (Science News, Vol. 144, July 24, 1993, pp. 55).

FUZZY LOGIC - AN INTRODUCTION - PART 3

Introduction

This is the third in a series of six articles intended to share information and experience in the realm of fuzzy logic (FL) and its application. This article and the three to follow will take a more detailed look at how FL works by walking through a simple example. Informational references are included at the end of this article for interested readers.

The Rule Matrix

In the last article the concept of linguistic variables was presented. The fuzzy parameters of error (command-feedback) and error-dot (rate-of-change-of-error) were modified by the adjectives "negative", "zero", and "positive". To picture this, imagine the simplest practical implementation, a 3-by-3 matrix. The columns represent "negative error", "zero error", and "positive error" inputs from left to right. The rows represent "negative", "zero", and "positive" "error-dot" input from top to bottom. This planar construct is called a rule matrix. It has two input conditions, "error" and "error-dot", and one output response conclusion (at the intersection of each row and column). In this case there are nine possible logical product (AND) output response conclusions.

Although not absolutely necessary, rule matrices usually have an odd number of rows and columns to accommodate a "zero" center row and column region. This may not be needed as long as the functions on either side of the center overlap somewhat and continuous dithering of the output is acceptable since the "zero" regions correspond to "no change" output responses the lack of this region will cause the system to continually hunt for "zero". It is also possible to have a different number of rows than columns. This occurs when numerous degrees of inputs are needed. The maximum number of possible rules is simply the product of the number of rows and columns, but definition of all of these rules may not be necessary since some input conditions may never occur in practical operation. The primary objective of this construct is to map out the universe of possible inputs while keeping the system sufficiently under control.

Starting the Process

The first step in implementing FL is to decide exactly what is to be controlled and how. For example, suppose we want to design a simple proportional temperature controller with an electric heating element and a variable-speed cooling fan. A positive signal output calls for 0-100 percent heat while a negative signal output calls for 0-100 percent cooling. Control is achieved through proper balance and control of these two active devices.

SIMPLE FL CONTROL SYSTEM

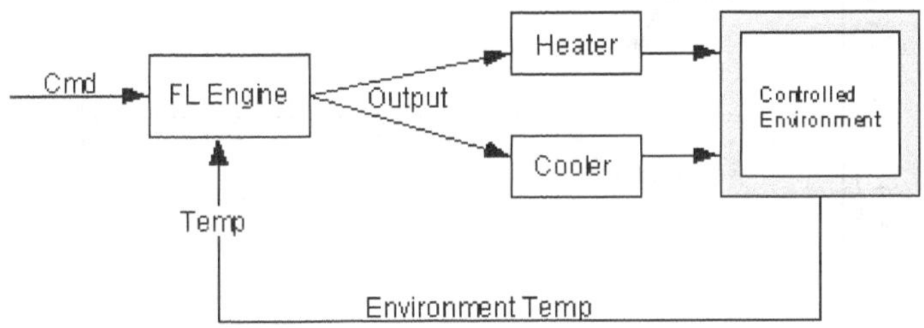

Cmd: Target temperature

Temp: Feedback sensor in controlled environment

Error: Cmd-Temp (+ = too cold – = too hot)

Error-dot: Time derivative or Error (+ = getting hotter – = getting cooler)

Output: HEAT or NO CHANGE or COOL

Figure A.1 A simple block diagram of the control system.

It is necessary to establish a meaningful system for representing the linguistic variables in the matrix. For this example, the following will be used:

"N" = "negative" error or error-dot input level

"Z" = "zero" error or error-dot input level

"P" = "positive" error or error-dot input level

"H" = "Heat" output response

"-" = "No Change" to current output

"C" = "Cool" output response

Define the minimum number of possible input product combinations and corresponding output response conclusions using these terms. For a three-by-three matrix with heating and cooling output responses, all nine rules will need to be defined. The conclusions to the rules with the linguistic variables associated with the output response for each rule are transferred to the matrix.

What is Being Controlled and How:

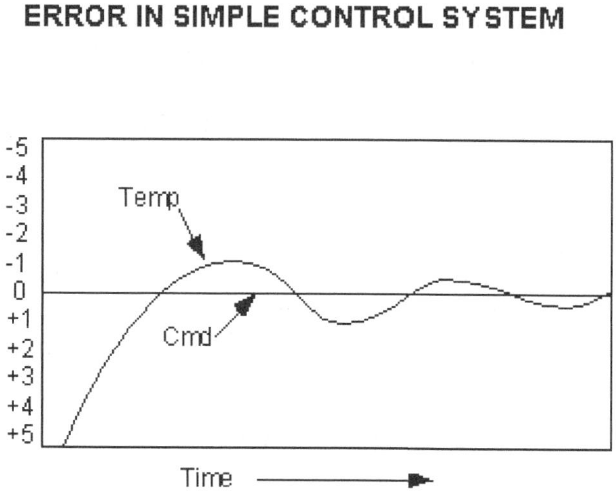

Figure A.2 Typical control system response

Figure A.2 shows what command and error look like in a typical control system relative to the command setpoint as the system hunts for stability. Definitions are also shown for this example.

Definitions

INPUT#1: ("Error", positive (P), zero (Z), negative (N))

INPUT#2: ("Error-dot", positive (P), zero (Z), negative (N))

CONCLUSION: ("Output", Heat (H), No Change (-), Cool (C))

INPUT#1 System Status

Error = Command-Feedback

P = Too cold, Z = Just right, N=Too hot

INPUT#2 System Status

Error-dot = d(Error)/dt

P = Getting hotter Z = Not changing N = Getting colder

OUTPUT Conclusion & System Response

Output H = Call for heating - = Don't change anything C = Call for cooling

System Operating Rules

Linguistic rules describing the control system consist of two parts; an antecedent block (between the IF and THEN) and a consequent block (following THEN). Depending on the system, it may not be necessary to evaluate every possible input combination (for 5-by-5 & up matrices) since some may rarely or never occur. By making this type of evaluation, usually done by an experienced operator, fewer rules can be evaluated, thus simplifying the processing logic and perhaps even improving the FL system performance.

RULE STRUCTURE & RULE MATRIX

Antecedent Block | Consequent Block

1. IF Cmd-Temp=N AND d(Cmd-Temp)/dt=N THEN Output=C
2. IF Cmd-Temp=Z AND d(Cmd-Temp)/dt=N THEN Output=H
3. IF Cmd-Temp=P AND d(Cmd-Temp)/dt=N THEN Output=H
4. IF Cmd-Temp=N AND d(Cmd-Temp)/dt=Z THEN Output=C
5. IF Cmd-Temp=Z AND d(Cmd-Temp)/dt=Z THEN Output=NC
6. IF Cmd-Temp=P AND d(Cmd-Temp)/dt=Z THEN Output=H
7. IF Cmd-Temp=N AND d(Cmd-Temp)/dt=P THEN Output=C
8. IF Cmd-Temp=Z AND d(Cmd-Temp)/dt=P THEN Output=C
9. IF Cmd-Temp=P AND d(Cmd-Temp)/dt=P THEN Output=H

Error - (Cmd-Temp)

Error-dot - (d(Cmd-Temp)/dt)	N	Z	P
N	1 C	2 H	3 H
Z	4 C	5 NC	6 H
P	7 C	8 C	9 H

Figures A.3 & A.4 The rule structure.

After transferring the conclusions from the nine rules to the matrix there is a noticeable symmetry to the matrix. This suggests (but doesn't guarantee) a reasonably well-behaved (linear) system. This implementation may prove to be too simplistic for some control problems, however it does illustrate the process. Additional degrees of error and error-dot may be included if the desired system response calls for this. This will increase the rulebase size and complexity but may also increase the quality of the control. Figure 4 shows the rule matrix derived from the previous rules.

SUMMARY

Linguistic variables are used to represent an FL system's operating parameters. The rule matrix is a simple graphical tool for mapping the FL control system rules. It accommodates two input variables and expresses their logical product (AND) as one output response variable. To use, define the system using plain-English rules based upon the inputs, decide appropriate output response conclusions, and load these into the rule matrix.

REFERENCES

1. "Fundamentals of Fuzzy Logic: Parts 1,2,3" by G. Anderson (SENSORS, March-May 1993).
2. "Fuzzy Logic Flowers in Japan" by D.G. Schartz & G.J. Klir (IEEE Spectrum, July 1992, pp. 32-35).
3. "Fuzzy Logic Makes Guesswork of Computer Control" by Gail M. Robinson (Design News, Vol. 47, Nov. 28, 1991, pp. 21).
4. "Fuzzy Logic Outperforms PID Controller" by P. Basehore (PCIM, March 1993).

FUZZY LOGIC - AN INTRODUCTION - PART 4

Introduction

This is the fourth in a series of six articles intended to share information and experience in the realm of fuzzy logic (FL) and its application. This article will continue the example from Part 3 by introducing membership functions and explaining how they work. The next two articles will examine FL inference and defuzzification processes and how they work. For further information, several general references are included at the end of this article.

Membership Functions

In the last article, the rule matrix was introduced and used. The next logical question is how to apply the rules. This leads into the next concept, the membership function.

The membership function is a graphical representation of the magnitude of participation of each input. It associates a weighting with each of the inputs that are processed, define functional overlap between inputs, and ultimately determines an output response. The rules use the input membership values as weighting factors to determine their influence on the fuzzy output sets of the final output conclusion. Once the functions are inferred, scaled, and combined, they are defuzzified into a crisp output which drives the system. There are different membership functions associated with each input and output response. Some features to note are:

SHAPE - triangular is common, but bell, trapezoidal, haversine and, exponential have been used. More complex functions are possible but require greater computing overhead to implement.. HEIGHT or magnitude (usually normalized to 1) WIDTH (of the base of function), SHOULDERING (locks height at maximum if an outer function. Shouldered functions evaluate as 1.0 past their center) CENTER points (center of the member function shape) OVERLAP (N&Z, Z&P, typically about 50% of width but can be less).

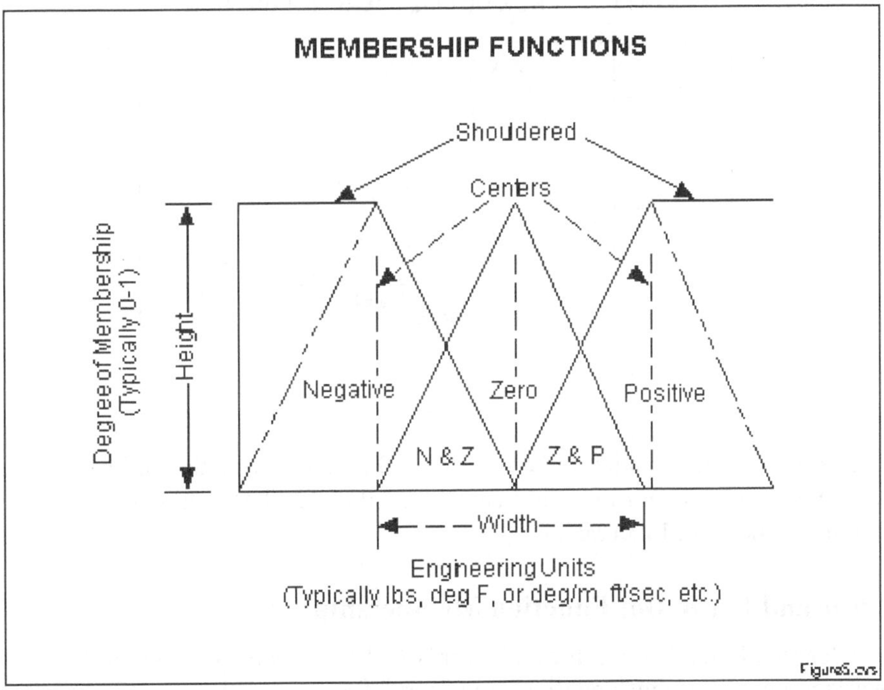

Figure A.5 The features of a membership function

Figure A.5 illustrates the features of the triangular membership function which is used in this example because of its mathematical simplicity. Other shapes can be used but the triangular shape lends itself to this illustration.

The degree of membership (DOM) is determined by plugging the selected input parameter (error or error-dot) into the horizontal axis and projecting vertically to the upper boundary of the membership function(s).

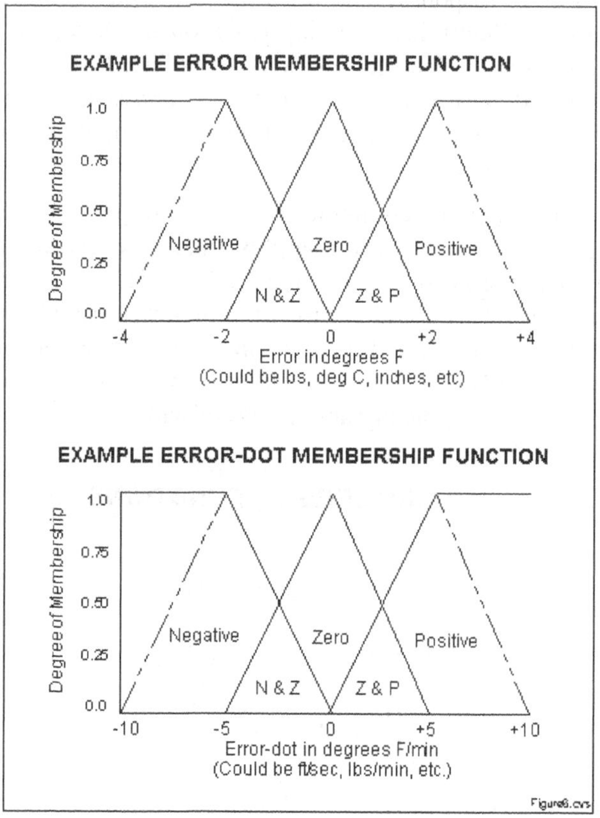

Figure A.6 A sample case.

In Figure A.6, consider an "error" of –1.0 and an "error-dot" of + 2.5. These particular input conditions indicate that the feedback has exceeded the command and is still increasing.

Error and Error-Dot Function Membership

The degree of membership for an "error" of –1.0 projects up to the middle of the overlapping part of the "negative" and "zero" function so the result is "negative" membership = 0.5 and "zero" membership = 0.5. Only rules associated with "negative" & "zero" error will actually apply to the output response. This selects only the left and middle columns of the rule matrix.

For an "error-dot" of +2.5, a "zero" and "positive" membership of 0.5 is indicated. This selects the middle and bottom rows of the rule matrix. By overlaying the two regions of the rule matrix, it can be seen that only the rules in the 2-by-2 square in the lower left corner (rules 4, 5, 7, 8) of the rules matrix will generate non-zero output conclusions. The others have a zero weighting due to the logical AND in the rules.

SUMMARY

There is a unique membership function associated with each input parameter. The membership functions associate a weighting factor with values of each input and the effective rules. These weighting factors determine the degree of influence or degree of membership (DOM) each active rule has. By computing the logical product of the membership weights for each active rule, a set of fuzzy output response magnitudes are produced. All that remains is to combine and defuzzify these output responses.

REFERENCES

1. "Fuzzy but Steady" (1991 Discover Awards) (Discover, Vol. 12, Dec. 1991, pp. 73).
2. "Neural Networks and Fuzzy Systems--A Dynamic Systems Approach to Machine Intelligence" by B. Kosko (Prentice-Hall, Englewood Cliffs, N.J., 1992).
3. "Putting Fuzzy Logic into Focus" by Janet J. Barron (Byte, Vol. 18, Apr. 1993, pp. 11).
4. "Putting Fuzzy Logic in Motion" by Dr. P. Miller (Motion Control, April 1993, pp. 42-44).

FUZZY LOGIC - AN INTRODUCTION - PART 5

Introduction

This is the fifth in a series of six articles intended to share information and experience in the realm of fuzzy logic (FL) and its application. This article will continue the tutorial discussion on FL by looking at the output membership function and several inference processes. The next article will wrap up the discussion of the ongoing example. To further explore the topic of FL, references are included for interested readers.

In the last article, we left off with the inference engine producing fuzzy output response magnitudes for each of the effective rules. These must be processed and combined in some manner to produce a single, crisp (defuzzified) output.

Putting it All Together

As inputs are received by the system, the rulebase is evaluated. The antecedent (IF X AND Y) blocks test the inputs and produce conclusions. The consequent (THEN Z) blocks of some rules are satisfied while others are not. The conclusions are combined to form logical sums. These conclusions feed into the inference process where each response output member function's firing strength (0 to 1) is determined.

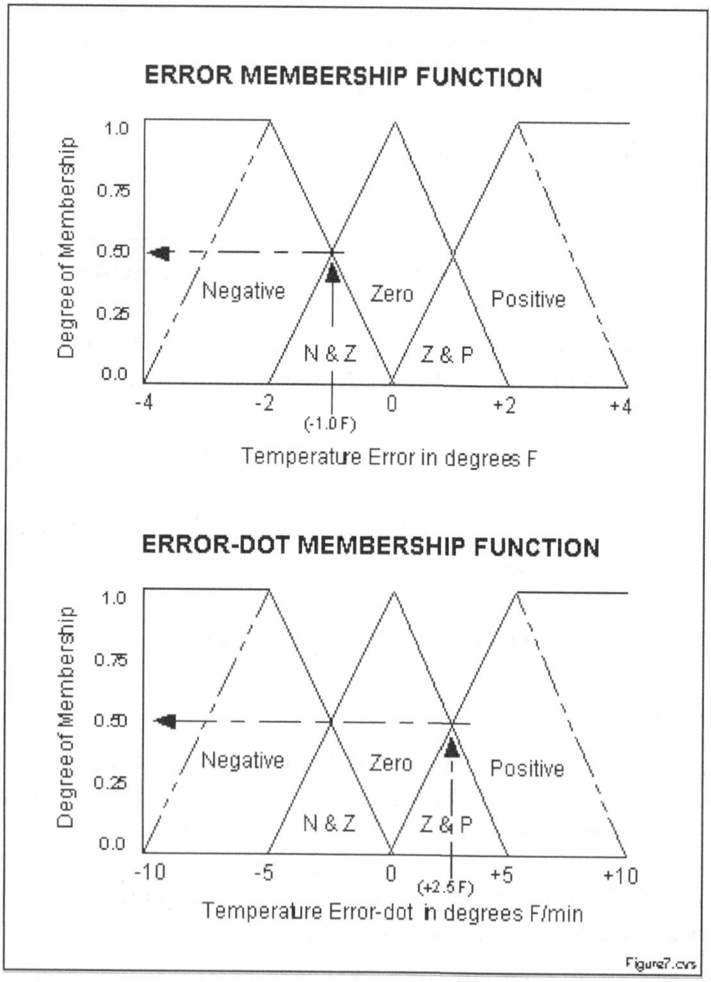

Figure A.7 Degree of membership for the error and error-dot functions in the current example.

Data summary from previous illustrations:

INPUT DEGREE OF MEMBERSHIP

"error" = –1.0: "negative" = 0.5 and "zero" = 0.5

"error-dot" = +2.5: "zero" = 0.5 and "positive" = 0.5

ANTECEDENT & CONSEQUENT BLOCKS (e = error, er = error-dot or error-rate)

Now referring back to the rules, plug in the membership function weights from above. "Error" selects rules 1, 2, 4, 5, 7, 8 while "error-dot" selects rules 4 through 9. "Error" and "error-dot" for all rules are combined to a logical product (LP or AND, that is the minimum of either term). Of the nine rules selected, only four (rules 4, 5, 7, 8) fire or have non-zero results. This leaves fuzzy output response magnitudes for only "Cooling" and "No_Change" which must be inferred, combined, and defuzzified to return the actual crisp output. In the rule list below, the following ddefinitions apply: (e)=error, (er)=error-dot.

1. If (e < 0) AND (er < 0) then Cool 0.5 & 0.0 = 0.0
2. If (e = 0) AND (er < 0) then Heat 0.5 & 0.0 = 0.0
3. If (e > 0) AND (er < 0) then Heat 0.0 & 0.0 = 0.0
4. If (e < 0) AND (er = 0) then Cool 0.5 & 0.5 = 0.5
5. If (e = 0) AND (er = 0) then No_Chng 0.5 & 0.5 = 0.5
6. If (e > 0) AND (er = 0) then Heat 0.0 & 0.5 = 0.0
7. If (e < 0) AND (er > 0) then Cool 0.5 & 0.5 = 0.5
8. If (e = 0) AND (er > 0) then Cool 0.5 & 0.5 = 0.5
9. If (e > 0) AND (er > 0) then Heat 0.0 & 0.5 = 0.0

SUMMARY

The inputs are combined logically using the AND operator to produce output response values for all expected inputs. The active conclusions are then combined into a logical sum for each membership function. A firing strength for each output membership function is computed. All that remains is to combine these logical sums in a defuzzification process to produce the crisp output.

REFERENCES

1. "Estimation of Fuzzy Membership from Histograms, Information Sciences" by B.B. Devi et al (Vol. 35, 1985, pp. 43-59).
2. "Fuzzy Logic" by Bart Kosko and Satoru Isaka (Scientific American, Vol. 269, July 1993, pp. 76).

3. "Fuzzy Sets, Uncertainty, and Information", by G.J. Klir and T.A. Folger (Prentice-Hall, Englewood Cliffs, N.J., 1988).

4. "Industrial Applications of Fuzzy Control" ed. M. Sugeno (North-Holland, New York, 1985).

FUZZY LOGIC - AN INTRODUCTION - PART 6

Introduction

This is the sixth and final article in a series intended to share information and experience in the realm of fuzzy logic (FL) and its application. This article will conclude the tutorial discussion of the ongoing FL example. For the interested reader, informational references are included.

Inferencing

The last step completed in the example in the last article was to determine the firing strength of each rule. It turned out that rules 4, 5, 7, and 8 each fired at 50% or 0.5 while rules 1, 2, 3, 6, and 9 did not fire at all (0% or 0.0). The logical products for each rule must be combined or inferred (max-min'd, max-dot'd, averaged, root-sum-squared, etc.) before being passed on to the defuzzification process for crisp output generation. Several inference methods exist.

The MAX-MIN method tests the magnitudes of each rule and selects the highest one. The horizontal coordinate of the "fuzzy centroid" of the area under that function is taken as the output. This method does not combine the effects of all applicable rules but does produce a continuous output function and is easy to implement.

The MAX-DOT or MAX-PRODUCT method scales each member function to fit under its respective peak value and takes the horizontal coordinate of the "fuzzy" centroid of the composite area under the function(s) as the output. Essentially, the member function(s) are shrunk so that their peak equals the magnitude of their respective function ("negative", "zero", and "positive"). This method combines the influence of all active rules and produces a smooth, continuous output.

The AVERAGING method is another approach that works but fails to give increased weighting to more rule votes per output member function. For example, if three "negative" rules fire, but only one "zero" rule does, averaging will not reflect this difference since both averages will equal 0.5. Each function is clipped at the average and the "fuzzy" centroid of the composite area is computed.

The ROOT-SUM-SQUARE (RSS) method combines the effects of all applicable rules, scales the functions at their respective magnitudes, and computes the "fuzzy" centroid of the composite area. This method is more complicated mathematically than other methods, but was selected for this example since it seemed to give the best weighted influence to all firing rules.

Defuzzification - Getting Back to Crisp Numbers

The RSS method was chosen to include all contributing rules since there are so few member functions associated with the inputs and outputs. For the ongoing example, an error of –1.0 and an error-dot of +2.5 selects regions of the "negative" and "zero" output membership functions. The respective output membership function strengths (range: 0-1) from the possible rules (R1-R9) are:

"negative" = (R1^2 + R4^2 + R7^2 + R8^2) (Cooling)

= (0.00^2 + 0.50^2 + 0.50^2 + 0.50^2)^.5 = 0.866

"zero" = (R5^2)^.5 = (0.50^2)^.5 (No Change) = 0.500

"positive" = (R2^2 + R3^2 + R6^2 + R9^2) (Heating)

= (0.00^2 + 0.00^2 + 0.00^2 + 0.00^2)^.5 = 0.000

A "Fuzzy Centroid" Algorithm

The defuzzification of the data into a crisp output is accomplished by combining the results of the inference process and then computing the "fuzzy centroid" of the area. The weighted strengths of each output member function are multiplied by their respective output membership function center points and summed. Finally, this area is divided by the sum of the weighted member function strengths and the result is taken as the crisp output. One feature to note is that since the zero center is at zero, any zero strength will automatically compute to zero. If the center of the zero function happened to be offset from zero (which is likely in a real system where heating and cooling effects are not perfectly equal), then this factor would have an influence.

(neg_center * neg_strength + zero_center * zero_strength + pos_center * pos_strength) = OUTPUT

(neg_strength + zero_strength + pos_strength)

$$\frac{(-100 \times 0.866 + 0 \times 0.500 + 100 \times 0.000)}{(0.866 + 0.500 + 0.000)} = 63.4\%$$

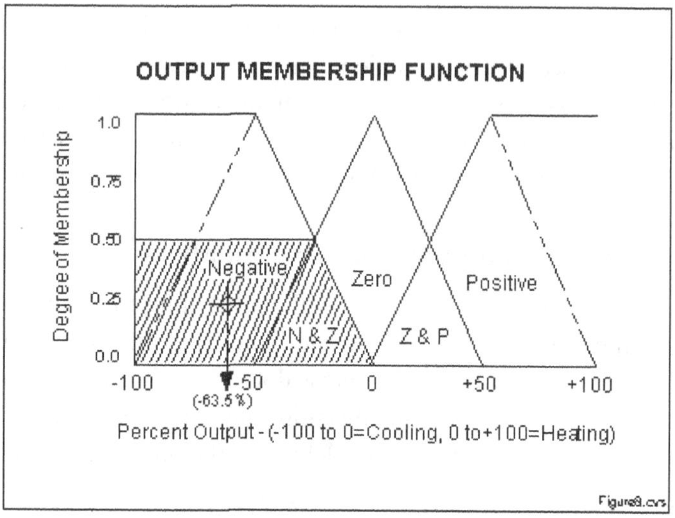

Figure A.8 The horizontal coordinate of the centeriod is taken as the crisp output.

The horizontal coordinate of the centroid of the area marked in Figure 8 is taken as the normalized, crisp output. This value of –63.4% (63.4% Cooling) seems logical since the particular input conditions (Error = –1, Error-dot = +2.5) indicate that the feedback has exceeded the command and is still increasing therefore cooling is the expected and required system response.

Tuning and System Enhancement

Tuning the system can be done by changing the rule antecedents or conclusions, changing the centers of the input and/or output membership functions, or adding additional degrees to the input and/or output functions such as "low", "medium", and "high" levels of "error", "error-dot", and output response. These new levels would generate additional rules and membership functions which would overlap with adjacent functions forming longer "mountain ranges" of functions and responses. The techniques for doing this systematically are a subject unto itself.

SUMMARY

The logical product of each rule is inferred to arrive at a combined magnitude for each output membership function. This can be done by max-min, max-dot, averaging, RSS, or other methods. Once inferred, the magnitudes are mapped into their respective output membership functions, delineating all or part of them. The "fuzzy centroid" of the composite area of the member functions is computed and the final result taken as the crisp output. Tuning the system amounts to "tweaking" the rules and membership function definition parameters to achieve acceptable system response.

CONCLUSION

This completes this article series on FL control and one way it can be done. The author has applied something close to this particular approach to a PC-based temperature controller which could be the topic of a future article series if there is interest. The PC solution has been implemented in Qbasic 1.0 and Borland's Turbo C running on the PC using iotech hardware ADC's and DAC's. This functionality has also been implemented using PIC's and 68HC11 processors.

Fuzzy Logic provides a completely different, unorthodox way to approach a control problem. This method focuses on what the system should do rather than trying to understand how it works. One can concentrate on solving the problem rather than trying to model the system mathematically, if that is even possible. This almost invariably leads to quicker, cheaper solutions. Once understood, this technology is not difficult to apply and the results are usually quite surprising and pleasing.

REFERENCES

1. "The Coming Age of Fuzzy Logic" Proceedings of the 1989 IFSA Congress, J.C. Bezdek, ed. (University of Washington, Seattle, WA 1989).
2. "The Current Mode Fuzzy Logic Integrated Circuits Fabricated by Standard CMOS Process" (IEEE Trans. on Computers, Vol. C-35, No. 2, pp. 161-7, February 1986).
3. "Fuzzy Logic - From Concept to Implementation", (Application Note EDU01V10-0193. ((c) 1993 by Aptronix, Inc, (408) 428-1888).
4. "Fuzzy Motor Controller" Huntington Technical Brief, D. Brubaker ed. (April 1992, No. 25, Menlo Park, CA 1992).

APPENDIX - B

SELF-ORGANIZING MAPS (SOM)

DATA CLUSTERING AND SELF-ORGANIZING FEATURE MAPS

Self-organizing feature maps (SOFM) - also called Kohonen feature maps - are a special kind of neural networks that can be used for clustering tasks. The goal of clustering is to reduce the amount of data by categorizing or grouping similar data items together.

A SOFM consists of two layers of neurons: an input layer and a so-called competition layer. The weights of the connections from the input neurons to a single neuron in the competition layer are interpreted as a reference vector in the input space. That is, a SOFM basically represents a set of vectors in the input space: one vector for each neuron in the competition layer.

A SOFM is trained with a method that is called competition learning: When an input pattern is presented to the network, that neuron in the competition layer is determined, the reference vector of which is closest to the input pattern. This neuron is called the winner neuron and it is the focal point of the weight changes. In pure competition only the weights of the connections leading to the winner neuron are changed. The changes are made in such a way that the reference vector represented by these weights is moved closer to the input pattern.

In SOFMs, however, not only the weights of the connections to the winner neuron are adapted. Rather, there is a neighborhood relation defined on the competition layer, which indicates which weights of other neurons should also be changed. This neighborhood relation is usually represented as a (usually two-dimensional) grid, the vertices of which are the neurons. This grid most often is rectangular or hexagonal. During the learning process, the weights of all neurons in the competition layer that lie within a certain radius around the winner neuron with respect to this grid are also adapted, although the strength of the adaption may depend on their distance from the winner neuron.

The effect of this learning method is that the grid, by which the neighborhood relation on the competition layer is defined, is "spread out" over the region of the input space that is covered by the training patterns.

Since SOFM learn a weight vector configuration without being told explicitly of the existence of clusters at the input, then it is said to undergo a process of self-organized or unsupervised learning. This is to be contrasted to supervised learning, such as the delta rule or back propagation where a desired output had to be supplied.

Clustering Methods

The goal of clustering is to reduce the amount of data by categorizing or grouping similar data items together. Clustering methods can be divided into two basic types: hierarchical and partitioned clustering. Within each of the types there exists a wealth of subtypes and different algorithms for finding the clusters.

Clustering can be used to reduce the amount of data and to induce a categorization. In exploratory data analysis, however, the categories have only limited value as such. The clusters should be illustrated somehow to aid in understanding what they are like. For example in the case of the K-means algorithm the centroids that represent the clusters are still high dimensional, and some additional illustration methods are needed for visualizing them.

Hierarchical Clustering

Hierarchical clustering proceeds successively by either merging smaller clusters into larger ones, or by splitting larger clusters. The clustering methods differ in the rule by which it is decided which two small clusters are merged or which large cluster is split. The end result of the algorithm is a tree of clusters called a dendrogram, which shows how the clusters are related. By cutting the dendrogram at a desired level a clustering of the data items into disjoint groups is obtained.

Partitional Clustering

Partitional clustering, on the other hand, attempts to directly decompose the data set into a set of disjoint clusters. The criterion function that the clustering algorithm tries to minimize may emphasize the local structure of the data, as by assigning clusters to peaks in the probability density function, or the global structure. Typically the global criteria involve minimizing some measure of dissimilarity in the samples within each cluster, while maximizing the dissimilarity of different clusters.

A commonly used partitional clustering method, K-means clustering will be discussed in some detail since it is closely related to the SOM algorithm.

THE K-MEANS CLUSTERING ALGORITHM

In K-means clustering the criterion function is the average squared distance of the data items u^k from their nearest cluster centroids,

$$E_K = \left\| \mathbf{u}^k - \mathbf{m}_{c(\mathbf{u}^k)} \right\|^2 \qquad \text{.....(B.1)}$$

where c (u^k) is the index of the centroid (mean of the cluster) that is closest to u^k.

One possible algorithm for minimizing the cost function begins by initializing a set of K cluster centroids denoted by m_i, i = 1…*K*. The positions of the m^i are then adjusted iteratively by first assigning the data samples to the nearest clusters and then recomputing the centroids. The iteration is stopped when E does not change markedly any more. In an alternative algorithm each randomly chosen sample is considered in succession, and the nearest centroid is updated.

The Algorithm

Suppose that we have *p* example feature vectors $u^i \in R^N$, i = 1..*P* and we know that they fall into *K* compact clusters, *K* < *P*. Let m^i be the mean of the vectors in Cluster-*i*. If the clusters are well separated, we can use a minimum-distance classifier to separate them. That is, we can say that u is in cluster C^k if $\| x - m^k \|$ is the minimum of all the k distances. This suggests the following algorithm for finding the *K* means:

THE K-MATRIX ALGORITHM

Given $\mathbf{u}^l \in R^N$, i =1..P

1. Make initial, i.e., t = 0, guesses for the means $\mathbf{m}^k(0)$ for cluster C^k, *k* = ..*K*

2. Use the means $\mathbf{m}^k$, $^k = 1..K$ to classify the examples $\mathbf{u}^i$, i = 1..N into cluster $C^k(t)$ such that

$$\mathbf{u}^i \in C^k \text{ where } k = \underset{j}{\arg\min} \left\| \mathbf{u}^i - \mathrm{m}^j \right\|$$

3. Replace $\mathbf{m}^k$ $k = 1..K$ with the mean of all of the examples for cluster C^k

4. Repeat steps 2 and 3 until there are no changes in any mean m^k, $k = 1..K$

$$\mathbf{m}^k(\mathrm{t}+1) = \frac{1}{\mathrm{card}\ (C^k(t))} \sum_{\mathbf{u}^i \in C^k(t)} \mathbf{u}^i$$

where card (C^k(t) is the cardinality of cluster C^k at iteration t, i.e., the number of elements in it.

In Figure B.1 it is shown how the means move into the centers of clusters

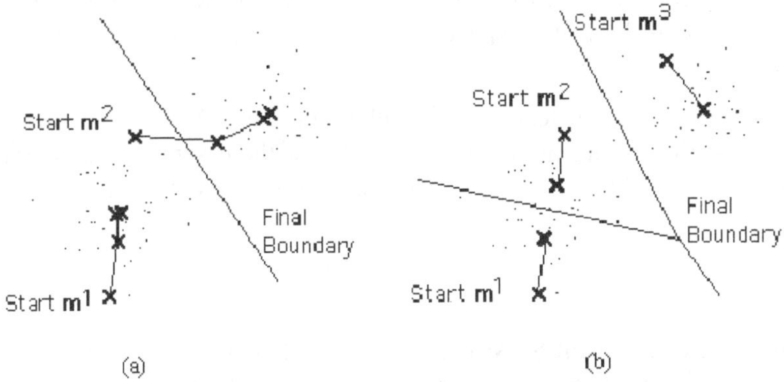

Figure B.1 Means of the clusters move to the center of the clusters as the algorithm iterates a) K=2 b) K=3

The results depend on the metric used to measure ||x - m^i||. A popular solution is to normalize each variable by its standard deviation, though this is not always desirable. A potential problem with the clustering methods is that the choice of the number of clusters may be critical: quite different kinds of clusters may emerge when K is changed.

Initialization of the centroids

Furthermore good initialization of the cluster centroids may also be crucial.. In the given algorithm, the way to initialize the means was not specified one popular way to start is to randomly choose *K* of the examples. The results produced depend on the initial values for the means, and it frequently happens that sub optimal partitions are found. The standard solution is to try a number of different starting points. It can happen that the set of examples closest to m^k is empty, so that m^k cannot be updated.

SELF ORGANIZING FEATURE MAPS

Self-Organizing Feature Maps (SOFM) also known as Kohonen maps or topographic maps were first introduced by von der Malsburg (1973) and in its present forms by Kohonen (1982).

SOM is a special neural network that accepts *N*-dimensional input vectors and maps them to the Kohonen layer, in which neurons are organized in an *L*-dimensional lattice (grid) representing the feature space. Such a lattice characterizes a relative position of neurons with regards to its neighbors, that is their topological properties rather than exact geometric locations. In practice, dimensionality of the feature space is often restricted by its visualization aspect and typically is $L = 1, 2$ or 3.

The objective of the learning algorithm for the SOFM neural networks is formation of the feature map, which captures of the essential characteristics of the N-dimensional input data and maps them on the typically 1-D or 2-D feature space.

Network structure

The structure of a typical SOM network for L=2 is shown in Figure B.2. It has *N* input nodes and *m*-by-*m* output nodes. Each output node *j* in the SOFM network has a connection from each input node *i* and w_{ij} denotes the connection weight between them.

During training the weights are updated according to the formula

$$w_{ij}(t+1) = w_{ij}(t) + \eta(t)(u_i - w_{ij})\|u_i - w_{ij}\| N(j,t) \qquad .....(B.2)$$

where w_{ij} and u_i are the ith component of the weight vector w_j of the neuron n_j, and the pattern u^k applied at the input layer is respectively, η (*t*) is the learning rate and *N* (*j, t*) is the neighborhood function which is changing in time

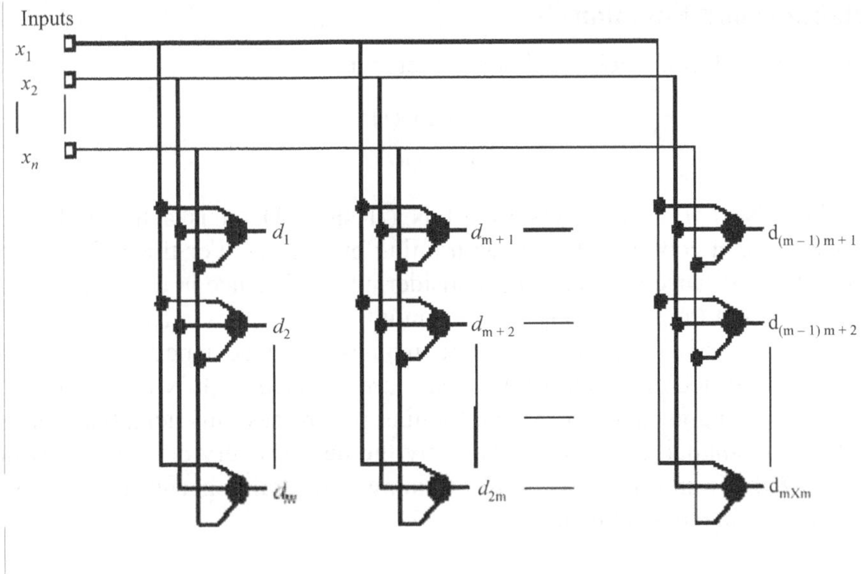

Figure B.2 Network topology of the SOM

The learning algorithm captures two essential aspects of the map formation, namely, competition and cooperation between neurons of the output lattice.

Competition

Competition determines the winning neuron d_{win}, whose weight vector is the one closest to the applied input vector. For this purpose the input vector u is compared with each weight vector w_j from the weight matrix W and the index of the winning neuron n_{win} is established considering the following formula.

$$n_{win} = \arg\min_j \left\| \mathbf{u} - \mathbf{w}^j \right\| \quad .....(B.3)$$

Cooperation

All neurons n_j located in a topological neighborhood of the winning neurons n_{win} will have their weights updated usually with a strength N (*j*) related to their distance d (*j*) from the winning neuron, where d (j) can be calculated using the formula

$$d(j) = \left\| pos(n_j) - pos(n_{win}) \right\| \quad .....(B.4)$$

where pos (.) is the position of the neuron in the lattice. As the norm city-block distance or Euclidian distance can be used.

Neighborhood Function

In its simplest form, a neighborhood is rectengular

$$N(j,t)=\begin{cases}1 & d(j)\leq D(t)\\ 0 & d(j)>D(t)\end{cases} \qquad \text{.....(B.5)}$$

Where N (*j*, *t*) is used instead of N (*j*) since D (*t*) is a threshold value decreased via a cooling schedule as training progresses. For this neighborhood function the distance is determined considering the distance in the lattice in each dimension, and the one having the maximum value is chosen as d (*j*). For L=2, N (*j*) corresponds to a square around n_{win} having side length=2D(*t*)+1. The weights of all neurons within this square are updated with N (j)=1, while the others remaining unchanged. As the training progresses, this neighborhood gets smaller and smaller, resulting in that only the neurons very close to the winner are updated towards the end of the training. The training end as remains no more neuron in the neighbourhood.

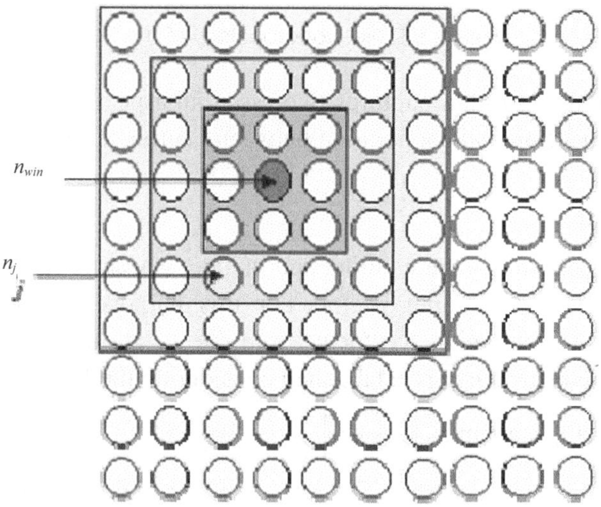

Figure B.3 Threshold neighbourhood, narrowing as training progresses. Usually, the neighbourhood function, N (*j*), is chosen as an L-dimensional Gausssian function:

Usually, the neighborhood function, N (*j*), is chosen as an L-dimensional Gaussian function:

$$N(j,t)=\exp\left(\frac{-d(j)^2}{2\sigma(t)^2}\right) \qquad \text{.....(B.6)}$$

where σ^2 is the variance parameter specifying the spread of the Gaussian function and it is decreasing as the training progresses as training progresses. Again σ is decreased Example of a 2-D Gaussian neighborhood function is given in Figure B.4.

Training SOFM

There are two phases of operation in SOM: the training phase and the clustering phase. In the training phase, the network finds an output node such that the Euclidean distance between the current input vector and the weight set connecting the input units to this output unit is minimum. This node is called the winner and its weights and the weights of the neighboring

Output units of the winner are updated so that the new weight set is closer to the current input vector. The effect of update for each unit is proportional to a neighborhood function, which depends on the unit's distance to the winner unit. This procedure is applied repeatedly for all input vectors until weights are stabilized. The choice of the neighborhood function, the learning rate, and the termination criteria are all problems dependent. The clustering phase is simple once the training phase is completed successfully. In this phase, after applying the input vector, only the winner unit is determined.

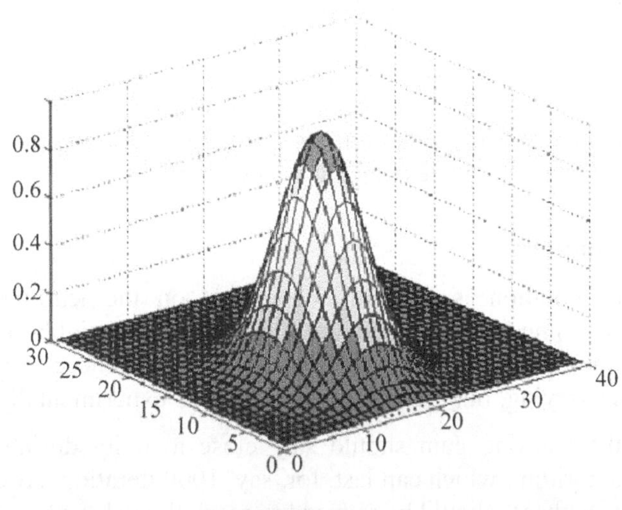

Figure B.4 2-D Gaussian neighborhood function for a 40 ×30 neuronal lattice

The training steps of SOM are as follows:

SOFT TRANING ALGORITHM

1. Assign small random values to weights $w_j = [w_{1j}, w_{2j}....w_{nj}]$;
2. Choose a vector $\mathbf{u}^k$ from the training set and apply it as input;
3. Find the wining output nodes n_{win} by the following criterion

$$n_{win} = \underset{j}{\arg\min} \left\| \mathbf{u} - \mathbf{w}^j \right\|$$

   where $\|.\|$ denotes the Euclidean norm and w_j is the weight vector connecting input nodes to the output node j;
4. Adjust the weight vectors according to the following update formula:

$$w_{ij}(t+1) = w_{ij}(t) + \eta(t)\,(u_i - w_{ij})\,N(j, t)$$

   where w_{ij} is the ith component of the weight vector $\mathbf{w}_j$, $\eta(t)$ is the learning rate and $N(j, t)$ is the neighborhood function;
5. Repeat steps 2 through 4 until no significant changes occur in the weights.

The learning rate $\eta(t)$ is a decaying function of time; it is kept large at the beginning of the training and decreased gradually as learning proceeds. The neighborhood function $N(jet)$ is a window centered on the winning unit n_{win} found in Step 3, whose radius decreases with time.

Neighborhood function determines the degree; an output neuron j participates in training. This function is chosen such that the magnitude of weight change decays with increase in distance of the neuron to the winner. This distance is calculated using the topology defined on the output layer of the network. Neighborhood function is usually chosen as rectangular, 2-dimensional Gaussian or Mexican hat windows.

Setting parameters

Feature map formation is critically dependent on the learning parameters, namely, the learning gain, η, and the spread of the neighborhood function specified for the Gaussian case by the variance, σ^2. In general, both parameters should be time varying, but their values are selected experimentally.

Usually, the learning gain should stay close to unity during the ordering phase of the algorithm, which can last, for, say, 1000 iteration. After that, during the convergence phase, should be reduced to reach the value of, say, 0.1.

The spread of the neighborhood function should initially include all neurons for any winning neuron and during the ordering phase should be slowly reduced to eventually include only a few neurons in the winner's neighborhood. During the convergence phase, the neighborhood function should include only the winning neuron.

Topological mapping

In SOFM the neurons are located on a discrete lattice. In training not only the winning neuron but also its neighbors on the lattice are allowed to learn. This is the reason why neighboring neurons gradually specialize to represent similar inputs, and the representations become ordered on the map lattice. As the training progresses, the winning unit and its neighbors adapt to represent the input even better by modifying their reference vectors towards the current input.

This topological map also reflects the underlying distribution of the input vectors as it is illustrated in the figures B.5 and B.6.

The sequence given in Fig. B.5 illustrates the feature space from its initial to final form as it is trained with the data, which is uniformly distributed in the square region. The dots are representing the data point used in the training. Neurons are organized in a 2-D lattice, their 2- D weight vectors forming an elastic string, which approximates two-dimensional object, which is a square. Each node is placed at the position corresponding to its weight. Initially the weights are assigned small random values, and cluster the data points around its weight vector. At the end, neighboring neurons clusters data points belonging to neighboring regions. Similarly the sequence given in Fig. B.6, represent formation of a 2-D feature map approximating a 2-D triangle from the input space.

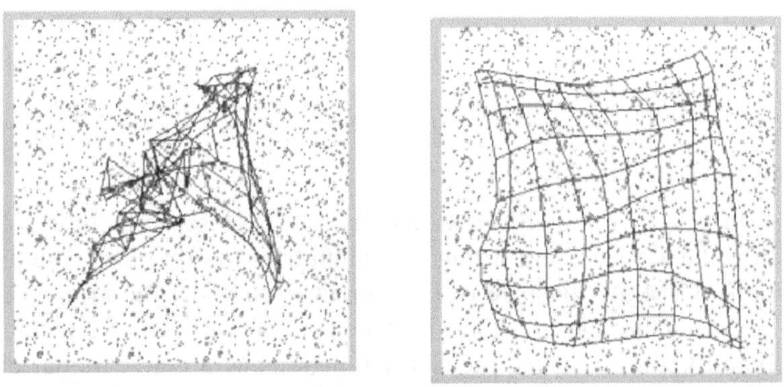

Figure B.5 2-D Feature map approximating a 2-D square from the input space

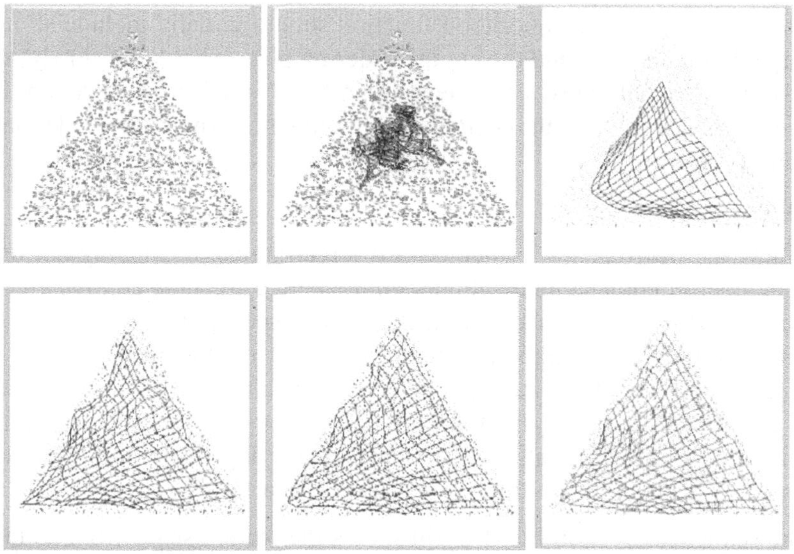

Figure B.6 2-D Feature map approximating a 2-D triangle from the input space

SOFM VERSUS K-MEANS CLUSTERING

Rigorous mathematical treatment of the SOM algorithm has turned out to be extremely difficult in general. In the case of a discrete data set and a fixed neighborhood kernel, however, there exists an error function for the SOM, namely

$$E = \sum_{k} \sum_{n_1 \in C_k} \alpha_{c_k} \left\| \mathbf{u}^i - \mathbf{m}_k \right\|^2 \qquad \text{......(B.7)}$$

The weight update rule of the SOM, corresponds to a gradient descent step in minimizing the above error function

Relation to K-means clustering

The cost function of the SOM, Equation (B.7), closely resembles Equation (2.1), which the K-means clustering algorithm tries to minimize. The difference is that in the SOM the distance of each input from all of the reference vectors instead of just the closest one is taken into account, weighted by the neighborhood kernel h. Thus, the SOM functions as a conventional clustering algorithm if the width of the neighborhood kernel is zero. The close relation between the SOM and the K-means clustering algorithm also hints at why the self-organized map follows rather closely the distribution of the data set in the input space: it is known for vector quantization that the density of the reference vectors approximates the density of the input vectors for high-dimensional data, and K-means is essentially equivalent to vector quantization. In fact, an expression for the density of the reference vectors of the SOM has been derived in the one-dimensional case; in the limit of a very wide neighborhood and a large number of reference vectors the density is proportional to $p(u)^{2/3}$, where p (u) is the probability density function of the input data.

APPENDIX - C

HOPFIELD NETWORKS

- The Hopfield Model
- Operation of the Hopfield Network
- Mathematics for the Hopfield Net
- Energy Landscape in Hopfield Networks
- The Hopfield Network as Content Addressable Memory
- Constraint Satisfaction/Optimisation using Hopfield Networks
- The Travelling Salesman problem

C. 1 The Hopfield Model

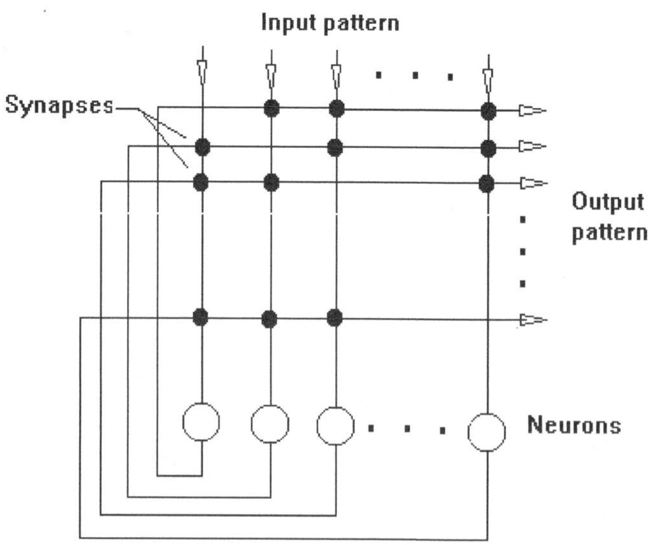

Fig. C 1 Architecture of the Hopfield network.

1. A fully connected network with binary (0/1, or +1/-1) inputs and outputs.
2. Symmetrically weighted ($w_{ij} = w_{ji}$).
3. Nodes perform weighted sum with a hard limiting (step) transfer function.
4. Output of each node fed back to the others.
5. Input applied to all nodes simultaneously and the network left to stabilise.
6. Outputs from the nodes in the stable state form the output of the network.
7. When presented with an input pattern, it outputs a stored pattern nearest to the presented pattern.
8. Good as content addressable memory and for solving optimisation problems.

C.2 Operation of the Hopfield Network

The Hopfield network has no learning algorithm as such. Patterns (or facts) are simply stored by setting weights to lower the network energy.

C.2.1 The teaching stage

The connection weights are set using the exemplar patterns from all classes according to the equation

$$w_{ij} = \begin{cases} \sum_{s=0}^{M-1} x_i^s x_j^s & i \neq 1 \\ 0 & i = j,\ 0 \leq i, j \leq N - 1 \end{cases}$$

where w_{ij} is the connection weight between node i and node j, x_i^S (either +1 or -1) is element i of the exemplar pattern for class s , and M is the number of pattern classes.

The result of the teaching stage is the association of a pattern with itself.

C.2.2 The recognition stage

The output of the net is forced to match that of the imposed unknown pattern.

$$\mu_i(0) = x_i \qquad 0 \leq i \leq N - 1$$

where $\mu_i(t)$ is the output of node i at time t.

The net is then allowed to iterate freely in discrete time steps until it converges (the output no longer changes).

$$\mu_i(t+1) = f_h\left[\sum_{j=0}^{N-1} w_{ij}\mu_j(t)\right] \quad i \neq j$$

The transfer function f_h is the step function.

The autoassociation of patterns means that presentation of a corrupt or incomplete input pattern will result in the reproduction of the original pattern as output. The network thus works as a content addressable memory (CAM).

C.3 Mathematics for the Hopfield Net

The connection strengths in a neural network can be represented by the elements of a matrix, with the subscripts i and j in weight w_{ij} (from node i to node j) corresponding to row and column numbers of the weight matrix W.

To form the weight matrix W for a Hopfield net, the outer product (also known as the cross product) of every pattern vector with itself is taken and then added up.

$$W = \sum_{M-1} \mathbf{x} \times \mathbf{x}$$

The outer product of two vectors is formed by multiplying each pair of vector terms.

For example, the outer product of (1 2 3 5) and (2 4 7 9) is:

	2	4	7	9
1	2	4	7	9
2	4	8	14	18
3	6	12	21	27
5	10	20	35	45

which forms a matrix. Such a matrix resulting from each pattern vector is added up to get the Hopfield net weight matrix W.

Alternatively, the two vectors can be considered as two matrices and a multiplication of the two matrices used to compute the weight matrix:

$$W = \sum_{M-1} x^T x$$

where x^T is the transpose of matrix x.

C.4 Energy Landscape in Hopfield Networks

The operation of the Hopfield network can be best understood in terms of the energy (error) function. The wells in the energy landscape represent patterns stored in the network. An unknown pattern represents a particular point in the energy landscape and as the network iterates its way to a solution, the point is expected to move towards one of these wells. These basins of attractions represent the stable states of the network.

A pattern can be stored in a Hopfield net by adjusting the weights such that when presented with that pattern, the energy will be lowered to correspond to one of the basins in the energy landscape.

Every time the state of the network changes, energy E either remains the same or falls,

$$\Delta E \leq 0 \qquad \text{.....(C.1)}$$

A change in state x_i of a neuron occurs only if

$x_i = -1$ and the activiation $(\sum_{i \neq j} w_{ij} x_j - T_i) > 0,\ \Delta x_i = 1 - (-1) = 2$

or

$x_i = 1$ and the activiation $(\sum_{i \neq j} w_{ij} x_j - T_i) < 0,\ \Delta x_i = -1 - 1 = -2$

So the product $\Delta x_i(\sum w_{ij} x_j - T_i)$ is always positive.

According to Hopfield (Eqn. (1))

$$\Delta E_i = \Delta x_i(\sum w_{ij} x_j - T_i)$$

$$E_i = x_i(\sum w_{ij} x_j - T_i)$$

$$= \sum_{i \neq j} w_{ij} x_i x_j - x_i T_i) \qquad \text{......(C.2)}$$

Thus the Hopfield net energy function *E,* obtained by summing (2) over all i, is a function of the weight values in the network as well as the patterns presented -

$$E = -1/2 \sum_{i} \sum_{j \neq i} w_{ij} x_i x_j + \sum_{i} x_i T_i$$

where w_{ij} represents the weight between nodes *i* and *j*,

x_i is the output of node *i,* and

T_i is the threshold value of node i.

In order to store a pattern, the value of the energy function for that particular pattern is minimised so that it occupies one of the minima in the energy landscape. Adding new patterns should not destroy the previously stored patterns although it will affect their storage to some extent.

Considering the energy function to minimise

$$E = -1/2 \sum_{i} \sum_{j \neq i} w_{ij} x_i x_j + \sum_{i} x_i T_i$$

The threshold value T_i can be set to zero, leaving the first term to be minimised. The weight matrix w_{ij} can be split into two parts w'_{ij}, which represents the effect of all the patterns except pattern s being stored, and w^s_{ij} the contribution made by pattern s alone.

$$E = -1/2 \sum_{i} \sum_{j \neq i} w'_{ij} x_i x_j - 1/2 \sum_{i} \sum_{j \neq i} w^s_{ij} x^s_i x^s_j$$

The first term cannot be altered directly but the second term can be minimised by making $\sum\sum w^s_{ij} x^s_i x^s_j$ as large as possible (it has a negative sign!).

To ensure that this term is always positive, we can equate

$$\sum_{i} \sum_{j \neq i} w^s_{ij} x^s_i x^s_j = \sum_{i} \sum_{j \neq i} x^2_i x^2_j$$

by making the weight term

$$w^s_{ij} = x^s_i x^s_j$$

Thus setting the values of the weights $w^s_{ij} = x^s_i x^s_j$ minimises the energy function for pattern s.

Updating values in a Hopfield network can be carried out in two different ways.

In *synchronous* updating, all nodes are updated simultaneously by freezing the network while the next value is computed across the network.

In *asynchronous* updating, a node is chosen at random and its output is updated according to the input it is receiving. The process is then repeated.

The difference between the two updating methods is that, in asynchronous updating, the change in output of one node affects the state of the system and therefore affects the next node's change.

Both updating methods should arrive at the same steady state, although by different paths.

C.5 The Hopfield Network as Content Addressable Memory(CAM)

A CAM stores a data item in such a way that it can be retrieved by presenting a key pattern (rather than its address) associated with it. Also called an *autoassociative* network, a CAM can be implemented using a Hopfield net. The main advantage is that the data item can be retrieved even when incomplete or corrupt information is presented.

In the Hopfield net, inputting incomplete information causes the network to react by following paths to a nearby energy minimum where the complete information is stored.

C.5.1 A simple example: A personnel filing system

A casting director needs a certain type of person - a brown-haired, blue-eyed young boy; other features do not matter.

A Hopfield net of 14 neurons is used. Characteristics of people are assigned to each neuron as follows:

1. male/female
2. brown/blue (eyes)
3. brown/blonde (hair)
4. thin/heavy
5. young/older (child)
6. young/older (adult)
7. through to 14) file number

A characteristic is given a value of +1 for the first choice, -1 for the second and 0 if neither is known.

Let the following three people be stored in the CAM:

1. Tom - brown hair, brown eyes, thin older child, file number 14
2. Bob - brown eyes, blond hair, heavy young child, file number 12
3. Sam - blue eyes, brown hair, heavy young adult, file number 7.

	1	2	3	4	5	6	7	8	9	10	11	12	13	14
Tom	1	1	1	1	-1	0	-1	-1	-1	-1	-1	-1	-1	1
Bob	1	1	-1	-1	1	0	-1	-1	-1	-1	-1	1	-1	-1
Sam	1	-1	1	-1	0	1	1	-1	-1	-1	-1	-1	-1	-1

The strength of each interconnection is determined from the activation values of the nodes being connected. E.g., in Tom's case nodes 1 and 4 are both active so the strength is increased, but in Bob's case, the values are opposite, so the connection strength is weakened. Every connection is adjusted as more people are added. The energy surface then will have many basins of attraction representing the stored facts.

The weight matrix **W** for the network is formed by taking the outer product of every fact vector with itself, and adding up all the products.

To retrieve information, partial data can be applied and the output neurons, after they have stabilised, will give the characteristic of a person closest to the input applied. The vector to be used as input in this example is **X** = [1 -1 1 0 1 0 0 0 0 0 0 0 0 0]

The weight matrix **W** and the transfer function *f* is applied to the input vector **X** to obtain a new vector **X'**

$$\mathbf{X'} = f(\mathbf{W} * \mathbf{X})$$

where '*' indicates inner product. **X'** becomes the input vector for the next iteration. This process is continued until the network converges (output stops changing).

The similarity between the input vector and the network state vector is measured by the Hamming distance, which gives the number of components that are different between the two vectors. In many situations, the network will settle to a pattern if this distance is less than half the length of the vectors.

C.6 Constraint Satisfaction/Optimisation using Hopfield Networks

C.6.1 Optimisation:

Picking the best answer out of many possibilities. Possible answers usually must meet one or more of constraints.

Examples:

Using a road map to travel between two suburbs

Locating chips in a circuit board with a given wiring layout

C.6.2 The travelling salesman problem (TSP) -

Given a set of n cities (A, B, C, ...) with distances d_{AB}, d_{BC}, d_{AC} etc., find a closed tour of all cities with a short total distance d.

A constraint satisfaction problem with the following constraints -

1. Each city to be visited once and only once
2. d to be as short as possible

Number of possible tours = n!

Number of distinct tours = n!/2n

Time required to find a solution with traditional methods is $O(t^n)$, i.e., increases exponentially with n - the so-called *np-complete* type problem.

If a neural net is used, adding a city will increase the amount of time required to find the solution, but will not increase the time exponentially.

C.7 Using the Hopfield net to solve the TSP

Inputs: Distances between cities.

The solution:

An ordered list of n cities, each city assigned one out of n possible positions.

n sets of n neurons used to represent a complete tour.

	1	2	3	4	5
A	0	1	0	0	0
B	0	0	0	1	0
C	1	0	0	0	0
D	0	0	0	0	1
E	0	0	1	0	0

The tour represented above is CAEBDC

To enable the network to compute a solution to the problem, an energy function must be defined for minimisation. Such an energy function can be written taking into account the problem constraints. If $V_{X,i}$ represents the the networks outputs, then subscript X represents a city name and subscript i represents the position it appears in the tour. The energy function E can be written as

$$E = A\sum_{X}\sum_{i}\sum_{j \neq i} V_{X,i}V_{X,j} + B\sum_{i}\sum_{X}\sum_{X \neq Y} V_{X,i}V_{Y,j}$$

$$+ C\left(\sum_{X}\sum_{i} V_{X,i} - n\right)^2 + D\sum_{X}\sum_{Y \neq X}\sum_{i} d_{X,Y}V_{X,i}(V_{Y,i+1} + V_{Y,i-1})$$

where A, B, C > 0.

1. Term 1 is 0 iff each city row contains no more than one 1, ie, no city appears more than once in any one tour.
2. Term 2 is 0 iff each tour position column contains no more than one 1, i.e., no two cities appear in the same position in any one tour.
3. Term 3 is 0 iff there are n entries of 1 in the entire matrix, ie, each tour must contain n cities.
4. Term 4 is minimised when the length of the tour is minimised. Measures the total distance by adding intercity distances $d_{X,Y}$ for each pair of adjacent cities.

1, 2 and 3 above are constraints to do with valid tours, and 4 is the constraint requiring the minimum length tour.

Combining the above, if *A, B* and *C* are sufficiently large, the low-energy states of the network will form a valid tour. The total energy of a valid state will be the length of a tour, and the states with shortest paths will be the lowest-energy states.

The connection strengths can be expressed in terms of the constants *A, B, C* and *D* (see [1]). In order not to bias the network towards any one tour, small random values are applied as input and the network is allowed to calculate an optimal result. Different starting values lead to different tours but they should all represent good solutions.

The most difficult problem is finding a suitable set of constants that guarantee a valid tour and allow the net to converge within a reasonable length of time. Although interesting from a theoretical point of view, the applicability of the TSP problem is limited. Also actual results obtained with the Hopfield net to solve this problem is less than encouraging.